D0911845

PARRY SOUND
LOGGING DAYS

PARRY SOUND
LOGGING DAYS

by JOHN MACFIE

The BOSTON
MILLS PRESS

A BOSTON MILLS PRESS BOOK

Copyright © John Macfie, 1987, 2003

Third printing, January 2003

Canadian Cataloguing in Publication Data

Macfie, John 1925–
   Parry Sound: logging days

ISBN 1-55046-055-2

1. Logging — Ontario — Parry Sound Region.
2. Loggers — Ontario — Parry Sound Region.
I. Title.

SD538.3.C3M33   1987   634.9'82'0971315   C87-094694-3

Published by
BOSTON MILLS PRESS
132 Main Street
Erin, Ontario, N0B 1T0
Tel. (519) 833-2407
Fax (519) 833-2195
books@bostonmillspress.com
www.bostonmillspress.com

IN CANADA:
Distributed by Firefly Books Ltd.
3680 Victoria Park Avenue
Toronto, Ontario M2H 3K1

IN THE UNITED STATES:
Distributed by Firefly Books (U.S.) Inc.
P.O. Box 1338, Ellicott Station
Buffalo, New York 14205

Cover painting by John Macfie
Cover design by Gill Stead
Printed by Ampersand, Guelph

The publisher acknowledges the financial support
of the Government of Canada through the
Book Publishing Industry Development Program
(BPIDP), for its publishing efforts.

CONTENTS

ACKNOWLEDGEMENTS

The Parry Sound Public Library kindly allowed me to use several photographs from their historical collection. Edna Knight of Parry Sound provided many photographs taken by her father, George E. Knight, as did James Isbester of Arnprior from his collection of photographs taken by his grandfather, James T. Emery. Several others gave me one or two photographs each, for which I am grateful. The Ontario Ministry of Natural Resources made an important contribution by allowing me to use excerpts from George Knight's recollections of river-driving, published originally in a 1950 issue of the Ministry's magazine, *Sylva*. I wanted Knight's words to be in the book, not only because several of his photographs appear in it, but because we were co-workers briefly; I joined the Ministry of Natural Resources as a young man not long before he retired from it. Thanks are also due to Aubrey Macdonald of Victoria, B.C., for letting me quote from the journal of his father, Duncan F. Macdonald, who at various times held the positions of Crown Timber Agent, Indian Agent and Homesteads Inspector in Parry Sound. The Parry Sound Public Library deserves further mention for giving me access to the voluminous Macdonald diaries, transcribed as a labour of love by former librarian Ray Smith. My daughter Beth critically reviewed my manuscript.

This book could not have been produced without the patient co-operation of half a hundred Parry Sound District citizens, nearly all of them now dead, who tolerated my intrusion in their living rooms and declining years with my tape recorder and probing questions. To them I owe the greatest debt of all. I dedicate this book to them.

John Macfie interviewing Nelson Clelland on the thresh floor of his barn.   — HEATHER BICKLE

PREFACE

When I was a boy in the 1930s, there was a general understanding in rural Parry Sound District that the next step in life upon leaving school was to "go to the bush." It had been so since the region was settled in the 1870s. A few radicals in each generation managed to break through to the outside world, but by and large two options were open: stay home on the farm, for whatever livelihood it might yield, or take up the life of a lumberjack. Because country families tended to be large and Parry Sound farms small, there was usually no choice at all for the younger sons.

The outbreak of world war in 1939 changed that. In a recruiting office a young man from Balsam, Hemlock or Sprucedale was as good as one from the city. When the Second World War ended, things did not return to the way there were — unlike following the First War. After three quarters of a century of exploitation the original stands of softwoods were gone, and a new era of small mechanized logging operations was beginning. The margin of overlap between the introduction of the power chainsaw to the Parry Sound woods and the disappearance of the lumber camp was remarkably narrow, five years at most.

While these events deflected my career in another direction, they did not extinguish my interest in the logging tradition, ignited by the tales of old shantymen and river-drivers living around Dunchurch, where I was born and raised. I began jotting down some of their stories in my teens. Then, when I got my hands on a tape recorder, I started systematically recording their reminiscences. This book is gleaned mainly from the more than one hundred reels and cassettes I accumulated from a quarter century of hunting down and taping old-timers. I must remind the reader that this is oral history, which is not always factual history. My informants had to reach back in memory, often very far. Oft-told tales mutate and gather embellishments, and some loggers, typical of outdoorsmen, are given to exaggeration — not to mention outright lying — when talking about their exploits. Still, there is a framework of fact in each story, and if it entertains as well as informs, so much the better. A great deal of repetition occurred in the tape transcripts, partly because the variety in a lumberjack's life was limited, and also, I suppose, because I tended to ask each interviewee the same questions. I have included some redundancies where I felt they added an interesting fact or shade of colour to the story. There are different approaches to breaking a logjam or doctoring a sick horse, and different ways to tell about it. Some informants dwell at length on the minutiae of logging, but I encouraged them to do so, to capture details of how things were done before they are lost.

The book's title might suggest it is a history of logging in the Parry Sound District, but considerable imbalance occurs in geographical coverage. The lower Magnetawan River area gets far more attention than, for example, the French River or eastern regions adjacent to Algonquin Park. This reflects the focus of my story-gathering efforts: the west-central parts of the district, where most of the old lumberjacks I knew lived. On the other hand, it occasionally strays outside the district's borders when a local logger found a season's employment in some distant place. The book therefore might be described more correctly as being about Parry Sound loggers.

Words and phrases peculiar to logging and river-driving, and sometimes to the Parry Sound woods, crop up frequently. Readers needing translation will find help in some drawings I have included and in the glossary at the back of the book.

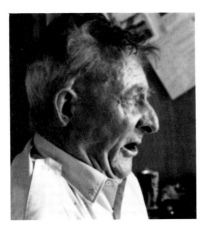

Jim McIntosh singing an old lumber camp ballad.

Walter Scott in 1982.

Gordon Whitmell (l.) and Marshall Dobson, c.1960.

Jack Campbell in the 1960s.

Rufus Harris in 1963.

CONTRIBUTORS

Walter Scott, Roy Smith, Jim McIntosh, George Beagan, Andy Ainslie, Jack Chisholm, William McKeown, Robert Hardie, Jack McAuliffe, Ernie Carlton, Tom Keating, Guy Smith, George Brunne, Dan Campbell, May Vowells, Gowan Gordon, Albert Scott, Les Wellington, Art Dobson, Hector Wye, Bill Scott, Rufus Harris, Roy Macfie, Marshall Dobson, Bernard Moulton, Roy Wainwright, Bert Little, Burley Harris, Mel Cameron, Norman Cameron, Willis Kenney, Everett Kirton, Bob Gibson, Jim Canning, Arnold McDonald, Jim Ludgate, Cliff Bennett, Fred Courvoisier, Gordon Whitmell, Roy Cochran, Alex Galipeau, Henry North, Levi North, Jack Campbell, Ed Pletzer, Dick Brear, Don Macfie, W.T. Lundy, Jim McArthur, Nelson Clelland, Jim McAmmond, Chris Watts, Bob McEachern, George Knight, Jim McAvoy, George Dobbs, Albert Bottrell, Arnold Madigan, Mike Giroux, Jack Laird, Marion Schell, Pearl MacLennan, William John Moore, Joe McEwen, Tony Green, Rupert Green, Duncan Campbell, Ethel North, Bert Currie, Bella Dickey, Rachel Irwin, Esther Einarson, Leo Madigan, Arthur Macfie.

Guy Smith being interviewed beside the Seguin River, 1980.

Jim Canning in 1963.

Albert Scott with a splitting axe and an Italian wine bottle, both relics of cordwood harvesting in Killbear Provincial Park.

PARRY SOUND DISTRICT
EARLY IN THE TWENTIETH CENTURY

TRUNK ROADS   – – – – –
RAILWAYS   +–+–+–+–+

Scale: 11/16" = 9 miles

1 – AHMIC HARBOUR
2 – BRUCE'S RAPIDS
3 – BURNT CHUTE
4 – BURK'S FALLS
5 – BURPEE SLIDE
6 – BYNG INLET
7 – CALLANDER
8 – CANAL RAPIDS
9 – CHAUDIERE DAM
10 – COMMANDA
11 – COPANANING
12 – DEPOT HARBOUR
13 – DILLON

14 – DOLLAR'S DAM
15 – DUNCHURCH
16 – FOURTEEN RAPIDS
17 – GOLDEN VALLEY
18 – THE GRAVES
19 – HORSESHOE FALLS
20 – HUNTSVILLE
21 – HURDVILLE DAM
22 – KEARNEY
23 – KIDD'S LANDING
24 – KILLBEAR POINT
25 – KNOEPFLI RAPIDS
26 – LOST CHANNEL

27 – LOVESICK RAPIDS
28 – MAGNETAWAN
29 – MAPLE ISLAND
30 – McKELLAR
31 – MOUNTAIN CHUTE
32 – MOUNTAIN RAPIDS
33 – NEEDLE EYE RAPIDS
34 – NORTH BAY
35 – PARRY SOUND
36 – PICKEREL
37 – PORT LORING
38 – POWASSAN
39 – ROSS' RAPIDS

40 – ROSSEAU
41 – SEGUIN FALLS
42 – SERPENT RAPIDS
43 – SOUTH RIVER
44 – SPRUCEDALE
45 – STEIDLER CREEK
46 – STURGEON FALLS
47 – SUNDRIDGE
48 – THREE SNYE DAM
49 – TROUT CREEK
50 – THE WILDCAT

INTRODUCTION

The tide of white pine logging, which rose in New England and Canada's Maritime provinces in the 18th century, reached Ontario's Parry Sound District in the latter half of the 19th century. Supplies of choice timber were running short on the Ottawa and St. Lawrence rivers, which flowed conveniently toward the main commerical centres, and settlement of the heart of the North American continent and the attendant building boom provided a natural market for the sawlogs dumped in Parry Sound's westward-running streams.

The beginning of logging in this area can be traced back to April 4, 1857, when cutting rights in a fifty-square-mile tract embracing the lower reaches of the Seguin River were awarded to William M. Gibson of Willowdale, Ontario. Gibson erected a water mill where the Seguin makes its final plunge into Georgian Bay and began sawing lumber. In the ensuing fifteen years timber limits taking in most of the district's best pine stands were allocated to lumbermen. By 1870 a steam-powered mill was in operation at the mouth of the Magnetawan River, followed by another at the main outlet of the French. Even greater quantities of sawlogs than these mills consumed were being towed down Georgian Bay to Midland, or across Lake Huron to Saginaw and Bay City in Michigan.

The pine era ended shortly after 1920, by which time nearly all the good stands had been cut down or destroyed in the devastating forest fires that inevitably followed clear-cut logging of resinous evergreens in a time when fire prevention and suppression capabilities were minimal. Even before 1900 lumbermen began to turn to the region's abundant supply of hemlock, primarily for bark, which was in demand for tanning leather, but also for its rather inferior lumber. Harvesting of hardwood species was phased in gradually early in this century as railways penetrated the area and inland sawmills became practicable (unlike pine, heavy maple and birch logs could not be floated down rivers to Georgian Bay), and within a decade or two pine had been relegated to a minor role in Parry Sound logging.

In contrast to more easterly parts of the white pine's range, Parry Sound played only a small part in the romantic square timber saga. Hewed in the bush from only the finest trees, those cumbersome sticks were taken all the way to Britain to be sawed into lumber, and it was a long way around the Great Lakes (or later by lumber baron J.R. Booth's railway from Ottawa to Depot Harbour) from Georgian Bay to the shipping point at Quebec. Instead, most trees were sectioned into sawlogs sixteen feet or less in length, piled in skidways until winter's snow lay deep, then hauled by horse-drawn sleigh to the nearest lake or river. During spring and summer the logs were floated downstream to Georgian Bay, some to be converted to lumber in sawmills located there, others to be corralled into booms and towed to distant mills. Throughout the open-water season a continuous parade of vessels carried away the output of local mills to deliver it down or across the Great Lakes. As a rule, both the winter's cutting and the summer's driving were done not by the companies holding cutting rights, but by "jobbers" who contracted the work. According to records kept at the Holland & Emery Company's inland depot at Ahmic Harbour, at the turn of the century the going rate for cutting and delivering pine to the Magnetawan River at Maple Island was $5 per thousand board feet. It cost the company an additional sixty-four cents a thousand to drive four million feet from Maple Island to Byng Inlet in the same year. According to one observer's estimate, nearly 200 million feet of pine came down the French River in one banner year.

Hemlock bark has a high tannin content (a substance which renders organic matter resistant to decay), so it was in demand before manufactured chemicals took over the process of leather-tanning. It took more bark than skin to make a given amount of leather, so tanneries were built close to the hemlock stands, in centres such as Parry Sound and Burks Falls, and cowhide which couldn't be bought locally was imported. Large quantities of bark were also shipped by rail to southern Ontario.

Sugar maples were then, as now, abundant in Parry Sound District and no finer specimens of yellow birch could be found anywhere. Maple and birch were in demand for flooring and interior wall-finishing, so as the land transportation network expanded, lumbermen turned to these species. The Second World War briefly placed the yellow birch on a lofty pedestal, for the skin of the Mosquito intruder aircraft, one thousand of which were made in Canada, was fabricated of yellow birch veneer.

By the 1930s nearly all tree species were being harvested for lumber, railway ties, shingles or pulpwood. Most of the ground then being logged had been gone over previously for pine. Still, life in the woods remained much the same as sixty years before. Trees were felled, sectioned into logs, piled on skidways and transported to the water or the mill by the muscle power of men and horses housed in rough camps, as always. Following the Second World War things changed rapidly, almost abruptly. Within five years trucks, bulldozers and tree-harvesting machines virtually eliminated the horse, and the power chainsaw replaced the crosscut saw.

Old-style logging was a labour-intensive industry, and the work force was drawn largely from the backwoods farms and villages of the district. This book tells the stories of some of those people, recalled many years later. Most of the stories were extracted from tape recordings made between 1962 and 1986, when the interviewees ranged in age from their sixties to middle nineties. Their combined memories embrace half a century of logging, from the 1890s, the high point of the pine era, to just before the internal combustion engine relegated the lumberjack to a minor role in the forest.

A great many logging companies were active in the region during the half century covered by this book, only a few of which are mentioned by the story-tellers. The average lumberjack was much more aware of the contractor or foreman — often one and the same person — under whose critical eye he built skidways or broke logjams, and of the roving woods superintendent, than of the corporation which ultimately profited or lost by his efforts, and this is reflected in the stories they told. Among these bosses, three names appear repeatedly: Dougald "Doog" Campbell, Albert McCallum and Samuel Ritter. I am not entirely sure why they, among scores, were etched so clearly in the memories of those I interviewed. It is true they exerted their authority mainly in central Parry Sound District, where I got many of my interviews, but so did many others whose names are forgotten. All three were extremely energetic men, but it would seem they also possessed a certain charisma which made them memorable. They were giants of a kind — though none were big men physically.

Albert McCallum was born in 1859 and turns up in folk history as a young man carrying the mail on his back from Dunchurch to the Ontario Lumber Company's inland depot deep in the wilderness west of Loring soon after that area was opened to settlement around 1880. Before 1900 he entered the field of contract logging north of the Magnetawan River, and just after the turn of the century he moved south to begin logging Hagerman Township for the Holland & Emery Lumber Company, later sold and renamed the Graves Bigwood Company. This proved to be a prosperous venture for McCallum, and other contracts with Graves Bigwood to the west in Burton Township followed. His last job was with the George Holt Timber Company in McKenzie Township, and this proved to be his undoing. The company refused to pay the full amount of a contract when low water stalled a log drive before it reached its destination. An ensuing lawsuit was taken all the way to the Supreme Court in England, where McCallum won, but the process cost most of the money he had accumulated in decades of logging. He died in Parry Sound in 1931.

Born in Scotland in 1860, Dougald Campbell emigrated to Canada with his parents, who in 1871 filed papers on a 200-acre homestead at Campbell Lake near Waubamik, eight miles up the Great North Road from Parry Sound. Dougald and three brothers — and in due course sons and nephews in considerable number — gravitated naturally into a logging industry which was active on all sides. In the first decade or two of this century it was a rare lumber camp or river-drive gang that couldn't boast a Campbell or two. Several Campbells rose above the rank of lowly lumberjack, but none matched Dougald as an all-round woodsman of renown. Not only was he widely admired, respected and liked, but he was successful in the rough-and-tumble world of lumbering when there were few safety nets to rescue a contractor who wrongly predicted a winter's snowfall or a summer's wind.

Never lacking in bush and river lore, Dougald gained needed help on the fiscal side of business when his daughter married Harry Atwell, an educated English remittance man. According to Campbell family lore, after contracting winter cuts and summer drives in the Seguin, Shawanaga, Magnetawan and Still River watersheds, Dougald retired comfortably well-off with thirty-two teams of horses and $33,000 in the bank. He died in 1933.

Unlike McCallum and Campbell, Ritter was a company man, brought in from Pennsylvania by the Dodge Lumber Company of New York to act as woods superintendent in their timber limits flanking the middle Magnetawan River. Based first at Byng Inlet, where Dodge's sawmill stood, then at McKellar, he finally settled in Ahmic Harbour to oversee the inland operations of Dodge and three or four companies which succeeded it, the last being Graves Bigwood and Company. As woods superintendent (or "walking boss," as the men in the woods and on the river invariably called a man in his position) his responsibilities were enormous. He planned and monitored the annual timber cut in several townships, set up company camps, negotiated cutting and log-driving contracts with private "jobbers," and distributed supplies and equipment from his warehouses on the waterfront in Ahmic Harbour. All of this he did with a staff of no more than two or three people. When Graves Bigwood ceased operations in the area, Ritter maintained his presence on the Magnetawan River with a much frequented summer cottage on Wawashkesh Lake, where his name is commemorated in Ritter's Narrows. He died in Toronto in 1937 at the age of eighty-seven.

The lumber company most frequently mentioned by the men is Graves, Bigwood and Company of Toronto, which operated a huge sawmill at Byng Inlet. During the first quarter of this century Graves, Bigwood held cutting rights in six or seven townships flanking the Magnetawan River below Burks Falls, the last in a succession of firms to do so, beginning with the Dodge Company of New York in 1869. The more prominent individual in the partnership was William E. Bigwood, a civil engineer born and educated in Vermont, who came to Canada before the turn of the century and settled in Toronto (a son was killed in action while serving with the Royal Flying Corps in France in 1917). A tall, athletic, handsome and friendly fellow, Bigwood made a lasting impression on his employees around the Byng Inlet mill, where he spent considerable time in the sawing season. When he died at age sixty three in 1927, the periodical *Canada Lumberman*, the bible of the business, was moved to declare that he was "probably the best-beloved man in the Canadian lumber trade."

Power house (built 1906) and falls, Parry Sound. Note the log flume in the middle.

A three man log-making crew consisting of two sawyers and a chopper begins to fell a white pine. — W.D. WATT/PAC PA 121795

Part One:
GOING TO THE CAMP

FIRST JOB

JACK CHISHOLM (b. 1896)

This Parry Sound District, when I was a boy, was practically a solid pine forest. There was one tree — Peter Lumber Company cut it — that was over seven feet on the stump. A white pine, up between Orrville and Seguin Falls. I went to work in the lumber bush when I was twelve years old, for Jack Watkinson at Seguin Falls. I was cutting trails to the logs so the team could get them out to put them on the skidway. When I started I got $26 a month. A ruling they had in the camps: if you quit in the middle of the month you got nothing for that month. But if they fired you in the middle of the month, they had to pay you a full month's wages. You were hired by the month, and that was a calendar month — Saturday, Sunday and every day of the week.

TOM KEATING (b. 1885)

The first winter I worked in the bush I was eleven years old. We took the pine off our 300 acres, put the logs out on Lorimer Lake. I drove the team from the last of October to the first of April. We drew most of them on a big wooden sloop. I'd drive out on the ice, take an axe and knock the hook loose, and the horses would jerk the sloop out from under the logs. My dad had a man working for him, paid him fifty cents a day. We had our own hay, oats and meat, and my mother was the cook, and when we wound up we'd made about $140. The logs came down by Manitowabing Lake to Parry Sound, [to] the Parry Sound Lumber Company.

I first started to go into the big lumber camps with Peter's. I had come to Parry Sound and worked at the William Beatty Company with Dick Bolton in the plumbing shop for a while. I didn't like that, so I got on the train at Rose Point and went down to Maple Lake, went into a camp on the first of September and came out on the the twentieth of March. The wages had come up then; we were getting about a dollar a day.

I worked in at Wilson Lake for Doog Campbell when I was seventeen years old. When I went there to work for McCallum, we cut through the same thing again, and cut trees bigger than we had cut before. That was where the 1913 fire went through, and we had to cut that burnt timber. McCallum's camp was at the mouth of Dogfish Creek. I brought the drive down Dogfish that year, and Joe Farley brought it down out of Simkoko. We ran in together. Joe took it and I was clerk.

That's where I lost my finger. We got a jam at the foot of a slide, and I was using a decking line with a [jam] dog on it, and a rope. I was holding the decking line for the man to back the horses up and hook on. I had some men on the jam too, and of course the chain was hanging in the swift water. A guy dumped a small log in off the jam and it went crosswise over the decking line and jerked my hand into the block. I came pretty near losing that hand. Just tied a rag around it and hit out. I was twelve miles from Ardbeg, and it was hot that day in June and just one drink of water on the road. I had three hours to catch the train, and I caught it. It was in the station when I got there. It's the only time I ever bummed my way. I never thought of money and didn't have any in my pocket [for a ticket]. The conductor came along and I said, "There it is. I got hurt and I had to trot twelve miles to catch you."

NORMAN CAMERON (b. 1894)
I went to the camp when I was thirteen years old, down at the Ess Narrows. Johnny Gadway was manager and Tom Bell was foreman. Johnny came from Saskatchewan, an old schoolteacher, a fine old fellow. That was for the Parry Sound Lumber Company. They towed the logs to Victoria Harbour. I cut trails for a teamster, old Jim Gulliver. How I come to get the job, Dad was cooking for Johnny, and Johnny used to go to Parry Sound once a month and bring the cash up in a little handbag. He'd fetch it in to our place and set it on the floor and stay all night, then go up to the camp next day, twelve miles the other side of Loring. He had a little team, Pretty and Ned. One would pace and the other would trot to beat the band. I always liked horses because Grandad got me a little horse to come to school in the summertime. Well, I used to tend old Johnny's team, currycomb them, and I'd ride halfway over here to Golden Valley with him. "Yeah," he said, "you're a good boy, come on in and we'll give you a job." Well, Dad was cooking there and he wanted me to stay home and go to school. And I wanted to go to work. So in I go to camp [carrying] a cotton bag, and Mother gave me a feather pillow to take — the first and I guess only feather pillow that ever was in a lumber camp in this country. A bunch of us young lads had a pillow fight about three days after I got there. We broke it up and had feathers all over the camp. I spent all the spare time I had for weeks gathering feathers up.

Dad said, "You shouldn't work over a week, or two at the outside, at a camp until you find out how much money you are going to get." When I came in from the bush at night I always went in through the cookery and out the other end to the washroom, and washed and hung up my cap, and I was all ready for supper. I came in this night and Dad says, "You should go and find out how much money." In I go to the office to see old Johnny. I says, "Dad says I should find out how much money I'm getting." The poor men were getting $14 a month and the good ones, the rollers and teamsters, were getting $26. So I asked Johnny how much he was going to give me. "I don't know," he says, "go and ask Tom Bell." (Tom Bell always stood at the stable door every night when the teamsters came in to examine every horse that went into the stable. If your horse was all cut up with snags or there were ridges on it where you had been whipping it with a gad, you lost your job as a teamster. That was Tom Bell. He was good to his horses, but they had to work. Feed 'em good and work 'em, but don't pound 'em, don't abuse 'em.) And I asked Tom. Tom said, "I didn't hire you, Gadway hired you. Go back and see Gadway. He can set your wages." I go back to see Gadway and he was sitting behind a long desk he had. In behind it he had boots and tobacco and stuff for the men. He was sitting in there with his spectacles on, and long whiskers, and I said, "Bell sent me back. He said you hired me, it's up to you to set my wages." "Well, I'll tell you now," he said. "If you'll be a good boy, I'll give you $26 a month." Well sir, I pretty near went through the roof of the office. Gosh, he's giving me the same as the good men he's got here. Was I ever tickled. I worked like Sam Hill. I earned that $26 a month. But I had good help. I worked cutting trails with an old bachelor by the name of Jim Finlayson, who died in his bed and all alone on the road going in to the Rainy Dam. That old fellow was a dandy axeman. He'd done chopping [when they chopped] the big pine logs off. He kept his axe in perfect shape and he ground mine too. He could cut a log off and you wouldn't see a mark of his axe. But brush, he hated brush. That was just fine for me, I could throw that brush. So I used to pile the brush out of his trail, and he says to me, "Any of those big logs that's crossways of your trail, just leave 'em there and I'll cut them." So we got along good. I stayed there till I got sick, got the flu in March and came home when the camp was just about done.

ROY COCHRAN (b. 1905)
The first [bush] work I did, I was only twelve years old. I worked in back of McKellar on the Balsam Road for a relation of mine and took out logs for Jimmy Taylor. At that time that's all there was doing, logging. Next fall, in 1918, I got in touch with Jack Campbell of McKellar and worked for him twelve miles back from the track at Drocourt. I put in nine years for Jack and Doog Campbell. We logged the far side of Deer Lake, Island Lake and Partridge Lake, all up through there. Those logs went down the Magnetawan. One drive was sawed at Byng Inlet and the rest went across to Penetang or Midland. McGibbon was one of the companies. We had a camp at Lost Channel for Schroeder one year too. In the fall, when the drive was over, I did the cadging.

The first year I was in the bush, I cut trails against grown men. I could cut more trails than grown-up men, and I got big wages for a boy. I got $35 a month, and $45 was high. I learned from the time I was six years old. My home was about a mile back off the Christie Road. When my mother died from typhoid fever I was seven years old, and Father was in the hospital. The day Mother died I was skidding wood from back on the farm to the house with a team of horses that weighed thirty hundred. Alone. I learned all this from [the time I was] a kid, and when I went to cut trails I understood how to do it. I've seen grown-up men, when they couldn't get logs in their trail, they'd come back in my trail to get logs because I had logs ready and they didn't. If Jack Campbell was standing here this minute he'd tell you those very words.

The next winter I went in for Doog Campbell. Doog Campbell was a great old fellow with boys. Wonderful, couldn't beat him. When I got working for him, nothing would do but I'd drive a team of horses. I knew horses perfect. I'd raked hay on the farm when I couldn't sit up on the seat because I couldn't reach the trip, and I'd have to get onto the trip with both feet because I wasn't heavy enough. Dougald gave me a team and I went into the bush, skidding from one gang to another, whenever it was long skidding. Then the next year I had a gang of my own.

ERNIE CARLTON (b. 1891)
I went to work when I was fourteen years old, for Charlie Harris back in behind Mary Jane Lake. Fourteen dollars a month, and you worked twenty-six days to get that. Cutting trails. The next year I worked in the same camp for Dick Robinson, for the Peter Lumber Company. Then I worked for Dick Cooper behind Eagle Lake. He had a job cutting hemlock for Graves Bigwood. I cut my foot there. He bandaged it up and carried me on his back from wherever his camp was on some little lake to down home. Two miles, I guess. At Christmastime it was pretty well healed up. We had a dance and I, like a damn fool, got up to dance and tore it apart. It never healed up till spring then. There's the scar yet.

I started cutting trails, then sawing and rolling. Sawing is one of the hardest jobs in the bush, because it's drag all the time. If you're handling a hook, you're lifting heavy lots of time, but you can walk around a bit. Still, it was a job you had to do. Billy Willard and I worked together two or three falls, rolled together, then loading. There were two winters Sammy Thompson was top-loader. He was one of the best I ever saw. When he'd get his outside logs on, we'd ask how many he wanted and he'd say, "How many you want to send." Just roll 'em in, tumble 'em in. He always seemed to have room up there. You very seldom had to get up and help him straighten a log around. A lot of top-loaders you did. A lot of top-loaders, they miss the pup and you gotta run around and throw it up to them again. I betcha he didn't miss that pup more than half a dozen times a day. You know, they have this pup on the end of a long chain. He'd throw the end of the chain down to us on rollers, to put around the logs we were sending up. We'd hand the pup to him, and he drove it into the log. When the horses pulled, he took his canthook and held that log until the other logs came up in the chain. We'd send three or four sometimes, but mostly one or two. As soon as those logs passed the pup, he'd reach over and grab it, and only a second to do it in, too.

CHRIS WATTS (b. 1911)

My first job was pulling in timber for the handyman building sleighs. Art Cooper was the blacksmith and Bill Barager the handyman. In about one week I went to the bush skidding logs. I got along real well and after about three weeks I was the youngest man to ever run a skidding gang.

It was surprising how high they could pile the logs. They'd have hardwood skids, five or six inches in diameter and whittled out to fit on the logs. These were bald skids, for the logs to go up on the skidway. And it was surprising how those guys could use their canthooks. I had one big French guy by the name of King, from Penetang, and Bert McRae. Could they ever use canthooks! If you had smaller logs sometimes they'd pull the line over four or five — choke them they called it — and the team would take them up in one chainful, just the same as one log, one man at each end with his canthook. If one end went ahead, one man would pull back on it and the other shove ahead. A single log would roll up. A group of logs would just slide up the skids. They'd hold the back log with the canthook, one on each end. They had a little snibby on the end of their canthook. One man would hit against the log and give a little shove and the other guy would give a little pull back on it. The hardest thing is when you get a bunch of eight foots and a bunch of sixteens. Naturally you'd like to have the long log behind the short log.

All the logs had to be cut with a six-inch stub-shot, so when they battered on the rocks in the rapids going to the mill they'd still have full-length boards. In other words, a sixteen-foot log had to be cut sixteen foot six inches. In those days the company wanted as many sixteen-foot logs as they could get. But sometimes, if you got a turned butt, you'd cut a fourteen or twelve off it. The logs we took out were eighty percent pine and twenty percent spruce. They had to be clear of rot. I remember we skidded logs with a little hole in the butt or a little rot in the centre from the skidway up a trail and left them there to rot. It was hard country to log on account of so many high mountains. I've seen us skidding logs three times from the top of a mountain — dump over a cliff, then go down and skid them again and dump over another cliff, then skid them to a skidway. What bothered me most, after they got the roads plowed the Forestry guys would come in and walk the roads, and if they saw a pine top they'd go in and raise heck with McRae. He had men out there cutting down trees that we were down in behind hills and there was no way you could get them with horses. Beautiful pine with eight or nine, maybe ten logs in them and from one foot to five feet across the butt. They should have left them in there, it was a shame to just cut them down and leave them.

ANDY AINSLIE (b. 1889)

Before I started to log, all the real big stuff was taken out for board timber. I was a small boy when Snakeskin was logged first. Johnston and Beveridge from Parry Sound logged there. So help me, they only took two logs out of a tree and left the rest lying there. You run onto tops of them yet. The man chopping, he chopped the trees down, then sometimes he butted them and sometimes the sawyers butted them. Their saws had sockets in the ends with sticks in them for handles, and they couldn't take the handle off the saw to get it out if the log would bind. They would have to pry it up to get the saw out.

I started working in the bush over the other side of Debow's, here in Hagerman, cutting trails for Dick Taylor, who jobbed for Graves Bigwood. I was thirteen. There was good white pine there. I couldn't look over any amount of the logs. It was all pine them days.

Large pine logs. — PARRY
SOUND PUBLIC LIBRARY

Cutting and skidding in Foley Township c.1885. Some of the finest examples of Parry Sound white pine occurred as scattered specimens in a predominately hardwood forest such as this.
— ONTARIO ARCHIVES

JACK McAULIFFE (b. 1901)

I was born in 1901 in the township of Ennismore, near Peterborough. It was known as the Holy Land. When I went up to Pakesley to go on the drive and went to put a chain on something, this foreman said, "Aw, shit, that's an Irishman's way of doing it!" He was from the Peterborough area and knew my accent. We lived on a farm and my father died when I was seven years old. There were two or three reasons why I went to a lumber camp. In 1918 I'd finished high school, a three-year course; you wrote middle school entrance into normal school, but you weren't accepted into normal school till you were eighteen. The only future seemed to be the farm, and I hated it. We were a little confused. If [the war] had gone on a year longer I would have been called up; you didn't know what direction to go. We never seemed to have any money, and we had been taught from boys that went to the logging camps and the river drives about the big stakes they had when they came out. We would listen to their tales of huge skidways and loads of logs, and adventure, and we wanted to go there and see this.

When lumberjacks came out after the winter, maybe they'd have $100 or $125. The booze [in Parry Sound] was in the Harbour; there were two or three hotels. These lumberjacks didn't mix their booze, they drank it straight and the liquor wasn't watered down, as it is today. When you went in they put the bottle up there and your glass was there — I guess it was ten cents a shot — and they drank to get drunk. They put them in a room and they were rolled during the night. Then they were broke and there was somebody to hire them in the morning to go back into the bush again.

That's the way we got on. Our fare from Peterborough was paid by old Ted Cavanaugh. He hung around the Montgomery House in Peterborough. By then it was dry; the Ontario Temperance came in in 1916. Men were scarce then and I guess Cavanaugh got a dollar or so for each man he hired. He got us a ticket for Pakesley and took us to the CPR station. I think if you stayed three months it wasn't deducted from your pay when you settled up. We went up on the Sudbury train. We had our supper at the boarding house at Pakesley, and I remember having blueberries, which was something new to me. There'd be acres of lumber piles at Pakesley. After supper they took us to Lost Channel in a Ford with flanged wheels that ran on the Key Valley Railroad. Schroeder's walking boss had a launch that took us to Camp Five at Ess Narrows. In the morning they showed us the trail, said, "There's your trail to Camp Two."

We landed in camp in the middle of September. There was another chap with me, a year older than I was, and he had been in the camp previously and knew the answers. The first thing the foreman said was "Can you hang an axe?" We said we could. I didn't know the first thing about hanging an axe, the other chap did. We got our axes hung and turned the old grindstone. A new axe, you know what sharpening it requires. We were with what they called the "buck beaver," building roads. In order for the roads to be ready for winter, all the trees that were in the way were taken out by the roots, and it was just as level as a paved road, ready for the snow and frost. we had to dig and chop around the tree stumps. It was a job the average young lad would say, "To hell with this."

Then we sawed. There were two sawyers and a chopper. The chopper notched the tree. He decided where he was going to fall it, and he measured it and cut the top off it. The chopper didn't trim it, that was done by the trail-cutters when they were skidding. The trail-cutter had to figure the best route for the horses. The guys in the sawing crews had a way of figuring out the best way to fall the trees to get the logs out without too much trail-cutting. The fellow I was sawing with knew about as much about sawing logs as I did. Some of the crews at Schroeder's, in that good pine, would put up 200 logs a day. We had to report each night at the van, as they called it, where the foreman and clerk stayed. The skidders had to report every night too.

You got to know a good saw. You went out with a freshly filed saw every morning. You rushed to get the saw you knew was a good one and you hid it under your bunk for the night.

After about two or three weeks the foreman said, "You're from a farm. Can you drive a

20

team?" I said, "Yes," so he gave me a team for a few days. They used a gin pole to load logs. Usually there were two teams skidding to the skidway, and this gin pole picked the logs up and placed them on the skidway. My job was, they had a team of horses on the gin pole and it was like loading hay into a mow with a hayfork. I'd go out a piece and pull, and when they'd yell I'd turn around and come back. I was quite thrilled to be given a team to drive. You were pretty proud if you had one, and me a kid.

RUFUS HARRIS
My father was blacksmithing for John Hammel in at Mary Jane Lake and they wanted a choreboy, so I went in there three weeks before Christmas. I was fifteen. I stayed till Christmas, then got homesick and gave her up, so my brother Burley went in and stayed till the camp broke up.

I worked a little while for McCallum, but I didn't like it. It was burnt timber, all black and dirty. Old Mac had a bunch of foreigners who could outwork us. They could pick up and carry small logs that were too dirty for us Canadians. And Mac would praise them for it.

On Saturday nights we would have a stag dance and sing songs and tell stories till twelve o'clock. Some of them would play cards in the blacksmith shop till Sunday morning. Some would have fifty pounds of tobacco and some of them would be broke. Outside of that, there wasn't much going on. We had to get up at five o'clock, get breakfast, get out in the· bush and have some work done before daylight, and we didn't feel much like kicking up our heels at night, coming in after dark. A buck a day. I was glad when Burley said, "Let's get out of here."

DON MACFIE (b. 1921)
Bob Farley was the first white baby born in Hagerman Township. When he was fourteen years old he caught a cadge sleigh in Dunchurch and went with it to a lumber camp around Loring, where he got a job as a choreboy. That night the men in the camp cut up so much that the boy Farley got scared and was afraid to go to bed. He ate his breakfast next morning in Dunchurch. Years later William Dobson moved down from Loring and became a close neighbour and fast friend of Farley. He was telling him about being in a camp where this boy came one day and got a job as choreboy, and who disappeared the same night. Said Mr. Farley, "And who you you think that boy was? Laws, it was me!"

JACK CAMPBELL (b. 1887)
I was born in Waubamik. The years that I should have been going to school I was working in the bush, and I've been in the woods ever since. At thirteen years old I started to cut trails, for the Beatty Estate. That's the job they put a fellow at when he's a boy. For cutting brush the average man had a three-and-one-half-pound axe, because you're wrists don't take it so good. You're trying to cut the gads, up from the ground. My father used to say, "Now lookit, my boy, you can't do that!" [Cutting down toward the ground.] He says, "As soon as your axe gets a little dull you gotta bend it over a stone to cut it. Cut up from the ground!" And of course you've got to watch, because it's easy to take a finger off, cutting up from the ground.

GORDON WHITMELL (b. 1899)
All I ever did in the bush was drive team. I was driving horses on the discs at home when I was eight years old. Dad couldn't lay the lines down or I'd have them. I quit school at Eastertime when I was in the Fourth Book. I could have passed the entrance that year if I'd went on. Dad wanted me to go on to school, but I didn't. We had the two teams then, and I could drive a team just as good as anybody. Dad jobbed the timber off the Cooper lot at the Jordan Creek the year I was fourteen, and I drove the team that winter too. I did some of the skidding and drew the bark to Ahmic Harbour.

I was sixteen the first winter I worked in the camp for McCallum in behind Ardbeg. McCallum had three camps back in there. Frank McCallum had a camp on Wilson Lake and Charlie Piattie had one out closer to Pointe au Baril. Joe Farley was running the camp I was at. Sometimes we had to walk pretty near an hour in the morning to the sleighs, and make five or six trips. We dumped on Dogfish Creek and we dumped some on Simkoko Creek. I had our team from home. I went in January. McCallum sent word down that he wanted another team, but old Auntie Stevenson was sick, and of course they raised Dad, and he wouldn't leave her. Dad didn't want me to go, but I struck off. I just took my clothes and a sleigh. She died in February and Dad got kind of worried about me, so he came up to the camp and stayed till the job was done. It was the twenty-third of March when we came out.

I mind one time I got scared pretty badly. I went to dump on Simkoko Creek and there was a bad sandhill. I wasn't loaded that heavy, but they just had green sand, they hadn't any burnt sand made there yet. I went down there pretty fast, and I don't know what happened, but the load pretty near all went off when I made the turn at the bottom. I had only three or four logs left to stand on. Another time the ice wasn't very good on Simkoko Creek and we were a little afraid of it. I jumped off the load one day to help dump it, and by God I went through the ice up to my waist. That was the last trip that night and I had to get the horses back to camp. I took my socks off and wrapped myself in feed bags and horse blankets. It was quite a life, but I liked it.

Rolling a log with a decking line, viewed from the top of the skidway. Senders stand at each end, ready to square it up on the skids with their canthooks. — GEORGE KNIGHT

BUILDING CAMP

GUY SMITH (b. 1885)

I ran a camp for the Conger one year, about four miles up the Moon River. I went in and built the camp and ran it till spring. When you are going to build a lumber camp, the first thing you've got to look for is water. You've got to have lots of water, be fairly close to a lake, then you find a place to build a camp. You pick a level piece of ground that has a slope, so everything runs away. Then you can start to build a road into it, and start cleaning up your land — an acre or acre and a half. Any tree that's going to reach the camp has to be cut down. You decide where you're going to build your cookery, where you're going to build your stables. You want your stables a couple of hundred feet from the sleep camp, maybe a little better. The cookery is built first, to get the cook going.

It doesn't take long to build where there's lots of timber. You take the big logs on the start, and keep putting logs on till you get it the height you want. You use a decking line, roll the logs up with horses, chop   round notches in the logs that're already up there, and the next one that comes up drops into them. The notches have to be in far enough that the logs come close together. You split wedges to drive in the cracks, then plaster over them — they gotta be wedged to hold the plaster. I built a camp once and I couldn't get lime. I used clay, mixed it with water and a bit of salt to hold it together, and heated it. The floor didn't go in till you got the sides up. A log floor, all adzed off. Then your roof goes on. You put cross beams across to hold the roof up, then put your rafters on. Go in a swamp and get a bunch of small spruce or balsam for the rafters. You use lumber and tarpaper for the roof.

At the same time, you're fixing the inside of the cookery to get the cook in as fast as you can. Benches are logs flattened off on one side. Bore holes to put in legs, a couple at this end, a couple at that end. A big round box stove sat in the middle of the sleep camp, maybe sixteen feet from the door. They used to have the stovepipe go away up in the air, then along to the far end and out, but now they just go up one or two pipes, then along, and they have racks on both sides for drying mitts and clothes. Now they use the heat.

All we put in a stable were eight teams. Any more and you'd have another stable; build them end to end, about twelve feet apart. You have a big enough door you can drive your team right through into your stall. There's a runway eight feet wide up the centre of the stable, made of timber adzed off and laid the long way, put in butt-and-top. You allow eight feet for each team of horses. It takes six feet for horses end to end, then you've got another couple of feet for the manger. And the stall is eight feet wide, with a floor of poles laid in endways. There's a wall made of small poles between each stall, and a good solid post with holes bored in it and pins to hang the harness on.

WALTER SCOTT (b. 1893)

You had to build camps. I was only in one place where there were camps built. They'd take off about ten percent for the camp building, but you can't build them for that. You want to get a nice level place. It takes quite a bit of room for stables, blacksmith shop, sleeping camps and cookery. We used to clean out an acre, anyway. I'd look for some place I could get water first, because it's an awful nuisance if you've got to lug water a long ways. Sometimes I'd dig a well up toward the sleeping camp and another one for the cook, so they wouldn't have far to carry water. You'd generally get some long spruce and put two or three in for stringers, then put two-by-sixes in for joists. Then put your floor on, then build your walls, put your two-by-fours up and board it in. Then they generally put building paper on first, then tarpaper over that and on the roof. Then lathe it about every foot so it wouldn't blow off. [In the sleeping camp] there weren't beds, just bunks built on both sides and across one end. We furnished ticks and they stuffed them with hay or straw. And we gave them a couple of blankets apiece, then maybe, when it got on in the winter time, we gave them an extra blanket.

Typical turn-of-the-century lumber camps. — GEORGE KNIGHT — EVERETT KIRTON

DRESSING FOR THE JOB: KNEE PANTS AND NO SOCKS

BOB GIBSON (b. circa 1892)

When my father was toting in the winter he never had any trouble with the cold. He never wore a sweater till he was sixty-five years old, and he just wore long leather boots and no socks — thick cowhide boots. One day this traveller came into the hotel at McKellar, and he had a big fur overcoat on and gauntlets. Mr. Manning said to him, "Surely it can't be all that cold. There'll be a man in here today with a felt hat on and leather boots with no socks in 'em." The traveller wouldn't believe it. He said, "If he comes, I'll treat the crowd, and if he don't come, you'll have to treat 'em." It wasn't long till my father pulled in, and they asked him, "Are you cold, Hughie?" "No, it isn't bad at all." "How many socks have you got on?" He whipped off his boot and showed them his bare foot. So the traveller told them, "Step up to the bar and have drinks before dinner."

BERNARD MOULTON (b. 1894)

Shantying was a hard life. You were lucky if you had a change of clothes — mackinaw pants, mackinaw coat, heavy socks and rubbers or shoepacs. A shoepac had no heel on it, built the same as a moccasin and slipperier than the mischief, made of tanned leather with about a four-inch top. Your feet wouldn't get wet in mild weather with shoepacs, but they would with moccasins. Shoepacs were warmer than rubbers, and deerhide moccasins were warmer still. Indians used to come into camps to sell deerhide moccasins. I bought a pair one time and they had quills off porcupines, dipped in something to colour them. I paid about $1.25 for them. They were good.

NORMAN CAMERON (b. 1894)

I worked in the camp before I wore long pants. I was fourteen going on fifteen when I got my first pair of long pants. I wore knee pants in camp, and long socks. Mrs. Sommacol in Arnstein knitted me a pair of long woollen socks for when snow got deep. They came above my knees and you tied a string around them. And you had Bird's mackinaw pants, just knee length, and a Bird's mackinaw coat, and that was your outfit. You could stand in wet snow all day, come in and give them a shake at night, hang them up and in fifteen minutes they'd be dry. They never got wet on the inside if you didn't rub them. They were made out of long wool, three inches long, and woven from the inside out. I wore rubbers or moccasins — moosehide or deerhide moccasins in real frosty weather — and socks over the outside of my pants. You could wade around in snow all day and no snow ever got up inside your pants. If you wore long pants it made a great big leg you couldn't tie your moccasins on, and it made your legs too hot. You put socks over long pants and pretty soon your legs would be burning and your feet freezing. Same on the drive — short pants, the shorter the better. If you fell in the water you didn't have so much water-soaked pants to pull out. It was easier to get up on a log with short pants than long ones, and the bottoms of your pants weren't getting all dirty and hooked up on something all the time.

You got mackinaw pants for $3.25. You could buy a pair of long Gendron shoepacs that came up to the knee — made by old Gendron from Penetang — and wade in water all day in them and not one drop would come through. I bought a pair of them from old man Schwartz at Commanda, an old German fellow who kept a store. I got my first outfit to go to the camp from him. I only had $27 I'd earned at seventy-five cents a day working on the old North Road as waterboy. I picked out everything I wanted: two suits of woollen underwear, these shoepacs, the pants, the coat, undershirt, outside shirt and a pair of leather mitts. I had it all piled up on the counter and Schwartz had it figured up to about twice what it would come to. I said, "No I've only got $27, and I've got to have that stuff or I can't go to camp." "Oh," he said, "I can't do nothing or I'd be losing money." I said, "OK, I'll have to leave it," and started for the door. He says, "Come on back. You can have it. I can lose money." But he didn't lose, he was still making a little bit. And that was my first outfit to go to a lumber camp when I was thirteen.

ARNOLD McDONALD (b. 1907)

When you went in you didn't draw wages. If there was anything needed you got it out of the van. Us fellows, we had our own socks and stuff, but fellows shipped in from outside, they'd have to get stockings, pants, shirts and stuff. We didn't have much van, so we'd have a bit of a stake when we came out. Of course we used to send money home too, write out a cheque — well, you'd give them a kind of order that went out to the mill, the Ardbeg office, and they'd send a cheque home to Mother. But other than that there was nothing for you to spend money on. In later years they got a washwoman in the camps, which was a good thing, because they always used to get lice in there. You paid a dollar a month for wash bill, and they paid her I don't know how much, and she washed your clothes for you.

JIM CANNING (b. 1872)

Hugh Gibson used to butcher cattle for the lumber camps. One cold day I met him at noon, cadging a sleigh-load of beef away up to Still River from his place in Maple Island. I was undressing my feet to get them warm and I had about four pair of socks on. Hughie pulled off his long leather boots with nothing but some straw in them. He said, "I put a bit of straw in 'em on cold days."

GORDON WHITMELL (b. 1899)

You had to be well clothed [on the sleigh haul]. Sometimes, if it was a level road you could get off and walk, and you could walk going back with the empty sleighs. Lots of days our feet got pretty cold. I wore a pair of horsehide moccasins in real cold weather. Then there were buckskin moccasins. And some wore overshoes with three or four pair of socks and an insole in them.

HENRY NORTH (b. 1903)   LEVI NORTH (b. 1898)

(The North brothers were interviewed together, and their individual voices could not be readily distinguished on the tape.)

Mr. Markham [school-teacher, homesteader and tailor at Maple Island] was an Englishman, quite a big man, he weighed 240. He didn't pretend to be doing much outside, just a bit of gardening. Markham's mackinaw was good. You'd find a picture of a bird on it someplace if you got a suit made, stamped on the inside. That was the brand, Bird's Wool, made in Bracebridge. That was before the First World War, when they learned how to spoil mackinaw. There's never been anything good since; they call it mackinaw, but it's not. The winter I was going in to Camp Eight, Nelson Clelland was going to Parry Sound so I asked him to pick me up a pair of new mackinaw pants. He bought a pair and I thought they were pretty good, they were kind of woolly. I got pawing through the bush and they were bare in no time. In two weeks I was darning them, and all I was doing was driving horses. They were just a bit of string they threw some fuzz on. And I've never seen good mackinaw since.

The one suit of Markham's I had, Dad got me that for $5, knee pants and a short coat. You could pour water on it and it would just run down, it wouldn't run through. I wore it two winters and the pants were getting pretty small. Frank Paul was in the camp and he didn't have any, just a pair of what they called "duck," so I gave him mine. I guess there was another winter in them yet without patching. I don't know if the coat ever wore out. It changed colour; it burned like a horse, got to be a kind of fox colour in the sun.

JIM McINTOSH (b. 1896)

A camp would hold a hundred men. We had the old-fashioned bunks — no bedsprings, an old straw tick. We'd go out to the stable and fill it full of hay when it got down. Lots of blankets, Bird's Woollen Mill in Bracebridge, they used to buy the very best of them woollen blankets from Bird's. Two big box stoves made out of barrels and lots of good wood. The choreboy would go around through the night firing them up. Nice and comfortable. They'd have an old man for choreboy. A young lad, he'd sleep in and forget to wake the teamsters. They used to pick an old man. He'd do a lot of sleeping through the day, and at night he'd be crawling around the camp nice and quiet. You'd never hear him. Then in the morning he'd have the lamps all lit when you went to get up. And the teamsters, he'd have their lanterns all lit waiting for them to pick up and go to the stable. He'd wake the teamsters up at a certain time every morning. Old George Pettigrew was choreboy for years. Another was old Sam McMillan. They were no good in the bush, too old, but they were good choreboys because they liked the job, and they were on the job and they knew what to do. They kept the camp all scrubbed and they used to do our washing, in a big tub with a washboard, and a big kettle for boiling water. The blankets were washed in the spring when the camp broke up. There'd be a bunch of men stay to start the log drive and they'd wash all the blankets. They'd hang up decking lines through the bush and put them out on a sunny day, double them up nice and pile them up on the tables for the next year.

Some of Albert McCallum's men in Hagerman Township about 1908. The three in the foreground wearing turtleneck sweaters are "Polacks," the loggers' name for anyone from Eastern Europe. Davey Black, the subject of a couple of anecdotes in this book, is behind the man with the full beard. — ADELINE TAYLOR

stock

snibby latch

circle

bill
or
head

J.M.

Canthook

Length about 3½ feet.
A peavey, used on the river
drive, was around 5 feet
long and had an iron tipped
nose like this

Roy and Frank Macfie dumping a sleigh load of pine logs.

TOOLS OF THE TRADE: SAWS, AXES, CANTHOOKS

ROY MACFIE (b. 1891)

I had to break my three brothers in on the [crosscut] saw, then all my sons. Teaching somebody to saw is the hardest job in the world.

JIM McINTOSH (b. 1896)

Maple Leaf was the leading saw. The Simonds saw was good. They were made in Hull. But the old Maple Leaf was the leader of all. To sharpen a saw you had to have a jointer to get your cutting teeth all the right length, and you had to have a gauge and a set, and good files. You could buy a file then that would do you a month. A file today, two or three filings and it's gone soft. You'd have your saw tipped toward you and file from the point back. Your jointer, you put your file in it and ran it along the full length of your saw, and when every tooth was touched you knew they were all the same length. Then you put your gauge on over your drag tooth, and if it stuck up a little bit you filed that down to your gauge. In frozen timber you've got to have the drags right up close [level with the cutting teeth], but in the summer they drop down, oh, I'd say the thickness of a tableknife lower. Then there was something else you have to learn. You take two men, if they're real heavy on the saw, putting lots of weight on it when they're sawing, you can cut those drags down a little bit more.

I filed eight saws a day. You carry a spare saw with you, and while you're filing, the gang uses the spare. A filing does just one day, although that depends on the men that are using it. You take in the summertime, when you fall a tree and it goes on the ground, a lot of them will cut right down into the gravel, maybe hit a stone. But the old-timers, no sir, they didn't saw that log right off if the tree lay on the ground. In the fall, hook the team onto it and roll her over and the trail-cutter, a couple of blows of the axe would finish cutting her off.

GUY SMITH (b. 1885)

A fellow that's chopping ahead of a saw doesn't grind his axe any more than he has to. The longer you can go without grinding an axe, the smoother it gets and the farther into the timber it will go. A good chopper will never grind his axe if he can get out of it, just give it a little rub with a fine whetstone. But road gangs and fellows that are limbing hemlock, they have to keep their axes well ground.

In a big camp there's a filer. He has a little shack with a place to file saws and a light overhead. There are usually five or six cutting gangs, and each gang has a number. The filer has a place where he hangs the saws, and if you are number one, your saw will be hanging there. When you come in at night, you take that saw down and put [the one used that day] up there. Those saws were to be filed next day. That's all the filer had to do. He had a good job. I filed in a camp one fall myself, but there were only three gangs cutting and I filed in the bush.

A crosscut saw has cutting teeth, and drag teeth to bring the shavings out. A saw with two cutting teeth [between each drag tooth] was for hardwood, and the one with four cutting teeth for softwood. You gotta keep those drags down the same length as the cutting teeth. If they're too long the cutting teeth can't get down deep enough. There's a drag gauge you set over the cutting teeth with a little notch that your drag shows up through if it's too long. You file the point down level and it will be the right length. Sometimes one cutting tooth gets longer than the others — some of them wear faster than others — and the teeth won't cut even because they ain't all the same length. So before you start filing you take a file and run it lightly along the teeth, and if a tooth is high it will take that off. You have a lever that grips the cutting teeth and bends them out. You can make a pretty narrow cut in hardwood, but for softwood you bend them out more to leave the teeth loose in the cut. There's nothing much to filing, as long as you don't file too heavy. Just sharpen it, that's all. Some saws were softer than others. You could tell a good saw from a bad saw by the ring of it. A good saw, there's a good ring to it when you're sawing.

ROY COCHRAN (b. 1905)

A lot of people were in the habit of [over] filing a canthook because they'd never put the hook on a log and seen whether the point was cutting out or sliding, to know which side to file. If it was cutting out, the outside of your bill was slanting in too much. If it was sliding, you filed the inside of the hook. If you filed your steel away and lost the weight of the head of your hook, finally it wouldn't catch nothing.

It takes twenty-two inches of iron to make the hook on a canthook. See, you have to beat the head back. Some of the older ones put twenty-four inches in, but their hook was too big and heavy. You heat it and put it over the anvil and bend it down. You let that end cool, put it in the fire with the back of it down and the point up, put it on the anvil, drive it back and jump it up, put it in the heading tool and shape it, draw the head out to a point and finish it off. Then you circle it. You make your latch, eight inches overall, bend it down and make your snibbies and measure back six inches to make the hole where the hook goes in. Then you drill in front of it and behind it for the bolts to go in. When the hook is closed the bill should point right between the snibbies. When they're made that way it doesn't matter if a log is six inches or thirty-six, it'll catch one just the same as the other.

Maple was mostly used for canthook stocks, or white ash. Oak was good, but oak would get warped; maple wouldn't do that. Soft maple made a beautiful canthook stock, but the rollers had to use a little common sense. The stock was light and it made a great hook for sending logs when they used the decking line. When they started loading logs on the sleighs we set the hooks a little different for the senders and top-loaders. The top-loader was catching logs and rolling them. The senders rolled them to a certain extent too, but when they'd go to send they [needed] a little more circle in the hook because they were catching the ends of logs all the time. If they didn't have enough bow in them the hook would hit the end of the log, bounce off and wouldn't catch. You could always tell if the bill of your hook was set right. You get it too wide and when you catch a log it will slide. And if it's turned in too far it will cut itself out, dig in and come out again. But if it's set right it will draw itself right in. Tongs were the same. When you open them wide the points should be not quite straight across, but a little bit out. Then when you put them on a log and draw they're forcing in.

An axe had to be in perfect shape. If I saw any stranger picking up my axe, I wasn't too long letting out a roar. He'd hit about two clouts with it, and I wouldn't have any face left on my axe because he wouldn't know how to use it. You must not chop backwards in a hemlock limb; if you do you'll break chunks out of [the axe]. You've got to limb from the butt toward the top. If you limb toward the top and never toward the butt, you can have an axe that chops like nobody's business and you won't break it. Some of them would go a week at a time and never touch their axes [with a grindstone]. They'd use a small whetstone.

When you're hanging an axe it's hard to get them perfect. It depends on the belly in the handle. But when your axe is hung perfect and you set the knob down here and the axe here, the axe will set right in the middle of the blade. One corner nor the other other won't hit. I make them out of maple and ironwood. You've got to have a stick big enough that will split down the middle and give you one handle on each side. You've got to make your axe handle with the grain. If you try to make it from the outside toward the centre it won't stand.

JACK CAMPBELL (b. 1887)

That's as long ago as I can remember, that they were taking square timber out at Waubamik. That's seventy-eight years ago [1896], because my father had a camp back at Nine Mile Lake and I remember my sister was born that winter. They drew the timber right from where it was made, out onto the Georgian Bay. There wasn't any railroad here in those days. They went in with the sleighs right beside the timbers and put them on. Put a chain around it and roll it up. They didn't want you to use canthooks, you'd make a mark and spoil timber. So it went from here to the Georgian Bay, and from there I don't know where.

They'd be thirty, maybe thirty-five feet, maybe more than that, whatever was in a tree. They didn't go up into the knots at all. That's why there was such an awful waste of timber. They'd knock a straight piece of bark off and they'd burn tag-alders, burn them black, and put the string around that, pull it, and it blackened it from one end to the other. Then they pulled it in the air and gave it a crack down on the white place on the side of the timber. Then a fellow jumped on top of it with his axe and score-hacked it all along, just in so that when he came along with the broad axe he took all those axe marks out. If a timber was too big they split off pieces about eighteen inches long, then score-hacked it after that again. A score-hacker would have an axe at least four pounds, and the handle was long.

Us young lads would be going along watching those old fellow that worked with the broad axes. They seemed to get a kind of pitch ahead of themselves, they were always stooped over hewing because the timber was laying down close to the ground. If you watch that fellow, he goes along with that broad axe and just hits the line. I declare you wouldn't think it possible, a man swinging one of those big broad axes over his head like that. Just leave one-half the line on each side of the axe. You'd think he wasn't even looking, he did it that easy.

ED PLETZER (b. 1887)

I've seen the blade of an axe pretty near fly to pieces from hitting a dry hemlock knot. I could hit an axe and tell whether it was soft or whether it was good. I was down at Seguin one time with a load of stuff, and Sam Kilgour was there and he had his axe. I says, "I wish I could get hold of an axe like that." He says, "When you're going home I'll have an axe out at the gate at my place for you." He brought me out a brand-new axe, and I used that axe for chopping when I was peeling bark. It was a tempered axe that wouldn't break and wouldn't get dull. I just carried a file and I'd file it once in a while.

Felling a pine near Parry Sound in 1887, before efficient crosscut saws took over the job from axes.
— PARRY SOUND PUBLIC LIBRARY

A round hook from a skidding chain bearing the hook identifying mark of a Croft Township logging contractor named Kennedy.

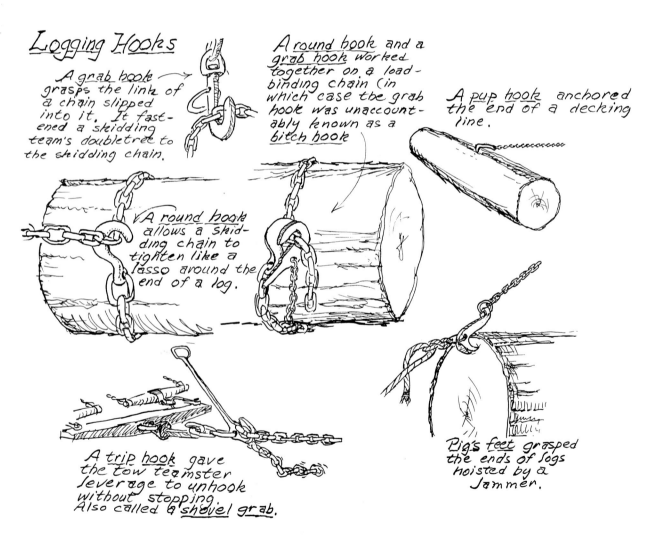

Logging Hooks

A *grab hook* grasps the link of a chain slipped into it. It fastened a skidding team's doubletree to the skidding chain.

A *round hook* and a *grab hook* worked together on a load-binding chain (in which case the grab hook was unaccountably known as a *bitch hook*)

A *pup hook* anchored the end of a decking line.

A *round hook* allows a skidding chain to tighten like a lasso around the end of a log.

A *trip hook* gave the tow teamster leverage to unhook without stopping. Also called a *shovel grab.*

Pig's *feet* grasped the ends of logs hoisted by a Jammer.

AT WORK
CHOPPING, LIMBING,
LOADING, SKIDDING

JACK CHISHOLM (b. 1896)

I worked at every job in the bush. I put in practically five years before I started the black-smith trade — sawing, cutting trails, skidding, rolling, driving team. I worked on the water tank and on the snowplow, and I drove the river. Probably you'd finish up in March haul-ing the logs out, then maybe it would be two or three weeks, and they'd start the drive. The drive would generally last till August, sometimes September, depends what luck you had getting through. Then you'd have two or three weeks and you'd be right back into the camp again, cutting logs. And while the drive was on they'd also put a gang in the bush cutting and peeling hemlock, 'cause they used that in the tanneries. So I knew every tool they needed, and I've made every tool they used in the bush — except these late years they're using jammers and stuff like that. Well, I used to build jammers at the last too.

The first skids were what they called "old grinners," spike skids. The senders rolled the logs up by hand, just with a canthook. Then they got the decking line and the guys learned how to send logs up a pair of smooth skids, "bald skids" they called them. You had to know how to handle them, because if you got one end a little too far ahead, she'd just go straight endways on the skidway, what they called "gunning." A loading [cant]hook, you make the head only a very little thicker than what the shank is, so if you get it pinched between two logs you can give it a little snap and out she comes. But the head, the other way, is deep. I put nine inches of one-inch steel in the head of the hook. I'd cut the steel twenty-one inches, then I'd bump her back till the shank was twelve inches. A lot of bumping.

They'd have an old experienced lumberjack as a sandpiper. Never see no young lads on a sand hill, because she's a tricky rig, knowing how much sand to put on. You can put too much, then your sleighs will start to jump and your whole load's liable to go off on top of the horses. If they start to slide, they're gone, especially the top logs. You see you put so many logs on, then you build what you call a cellar in the centre. Then you put your wrap-per on and drop your top logs in, and they bind it. The man on the sand hill was always called the sandpiper. The man giping, fixing the road, they called him a "chickadee." A top-loader was called a "sky-hooker" and the fellows sending to the top-loader were called "ground moles."

JIM McINTOSH (b. 1896)

I won a prize of $50 for skidding the most logs in one day, in Longford Township on Friday, September 5, 1928. All hemlock. That was a terrible township for timber. One year Jack McCracken was foreman, he got 175,000 pieces of hemlock. That's a lot of timber. It was all peeled hemlock. Every man had his day. Any team had a chance to do that any day of the fall he liked. And when you were going to have your day you went to the office and told the foreman, "I'm going to try it tomorrow." You had to skid them yourself. He sent a man out to see that nobody hooked that chain, and nobody unhooked it, the trail-cutters or any-body. Nobody swung your doubletree, you had to do that yourself. You've got two trail-cut-ters in the bush and two rollers. They could undo your chain when you came in if you wanted them to, and out in the bush the trail-cutter could swing your doubletree. But this man was sent out to see that nobody swung your doubletree and nobody hooked your chain. It was fair and square. The logs were piled four and five tier deep all around the skidway. I would say the longest draw would be 200 yards. Five, six and seven every chain load. Your skidding chain is twelve feet long, a round hook on each end. There'd be four logs piled up there and three here. Put that end around them three and this one around this four, and pull 'em together and away you go.

I had 375 to beat. When I went into the office that night Mr. Cooper, the general manager, said, "Are those all on one skidway?" I said, "Yes, they are, Mr. Cooper." He said, "In the morning, don't put another log on there, go ahead someplace else. I've got to go out and count those officially so everybody will know that there's no mistake made and I'm not trying to favour you." So him and two more men went out and counted them, 532. I got $50 that night and wages were $12 a month. That was for Standard Chemical, up in Longford Township, on the Black River.

GEORGE BEAGAN (b. 1890)

I mostly rolled. I was like your dad. Your dad was the best canthook man in Canada. He was left-handed. The first year that he rolled he loaded with Joe Farley and Ben French; they were the two that were considered the best men up in this North Country. They figured Farley and French were pretty good, but they hadn't nothing on your dad. I got to be pretty good myself, but I wasn't as good as your dad.

We had to send with the decking lines them times. There wasn't no jammer. Anybody at all can load logs now. But, by gosh, if you have to send them logs! There was a [canthook man] on each end of the skidway. If the fellows knew how to chain there wasn't much difficulty. You get a log with a big butt, taper down maybe eight inches to the top. Now there's generally an eight-foot mark on it. You want to get back a foot and a half or two feet, depends on how big that butt is, to get her balanced. There wasn't much to it if you knew how to chain. The senders chained turn about; one fellow chained for one load, the other fellow chained for the other. And that's the time you'd hear the swearing. You see, some fellows never got onto chaining the logs. Joe Shaughnessy and Art Ferris loaded in a gang opposite me one winter, and they were tearin' and jumpin' and roarin' and yellin'. We had got our three teams out and I said, "What's the matter, boys, you haven't got your three loads on?" "Oh, just a little trouble here." And the top-loader threw the chain down, and Shaughnessy threw it around this log. It was a bad one and I said, "You haven't got that chained right, Joe." And he said, "We can send anything with two ends on it!" It just started up the skids and went endways. You gotta have a little brains, too, as well as know how to handle a hook.

Oh, I wish you had seen that Hagerman limit when I first saw it. That was the most beautiful sight I ever saw. Get up on that mountain at the upper end of Shawanaga Lake and you could just look out over pine for miles and miles. There was no extra-big timber on that limit. There was only a very small percentage of big pine that would make board timber. You see, before Graves Bigwood got that limit, the Dodge I guess was first, then Holland and Emery — it was handed down — they started cutting board timber in this country, and they'd give some of the farmers a job of cutting five or six sticks. But them old fellows didn't know where the lines were and they didn't care. I was away up Steidler Creek and I came home and told my dad, I said, "I see where some Indian years ago cut down a great big tree, and the chips are all there." I says," I guess he's made a canoe out of it." And he says, "I guess I was the Indian." He'd made square timber. About three miles north of Shawanaga Lake there was a stand of old, old pine and he got a couple of trees and hewed them. He had an old black horse and drew them across onto Bolger Lake and across onto Snake Lake, shot it down over a hill off of Snake Lake and across onto Shawanaga Lake. And finally they had to go to Lorimer Lake. See, they were driven down the Seguin, then I guess went to the Old Country. Board timber sixty feet long. It didn't matter if it had a little sweep in it. It was loaded in boats with the sweep up, and as you kept putting the load up, it was all straight when it got over.

This old fellow was quite a one for the hotels, and Sam Ritter happened to come to the hotel and he told Ritter that Dan Beagan was cutting some pine on his place, three miles back [in the bush]. Ritter came down and he was going to hang my dad at sunrise for taking his timber. That was an organized township. There was a road allowance for sideroads, and they let that damn lumber company take timber off the road allowance. It was reserved. And when Ritter came to my dad and asked where he was taking the timber off, my dad took a flying guess and told Ritter what concession it was, the road allowance. "A

This photograph reveals how sleigh loads of logs were held in place. Four corner-bind chains fastened the outside logs of the bottom tier to the sleigh's bunks, and an encircling chain, tightened by the weight of a few logs placed on top, held the mass together. — GEORGE KNIGHT

Fastening Down a Load

The chains that 'wrapped' a sleigh load of logs in place were held tight by a few logs riding on top of them.

Tension could also be applied by means of an ironwood springpole. Used to bind loads of lumber and tanbark, and small loads of logs.

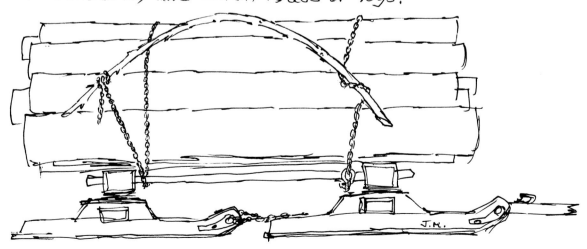

35

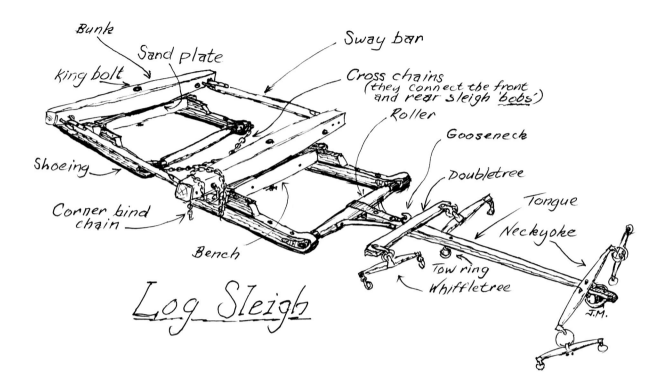

Bunk

Sand plate

King bolt

Sway bar

Cross chains
(they connect the front
and rear sleigh 'bobs')

Roller

Gooseneck

Doubletree

Shoeing

Tongue

Neckyoke

Corner bind
chain

Tow ring

Whiffletree

Bench

J.M.

Log Sleigh

The log dump on a Graves, Bigwood Company operation. — RENA SAAD

farmer," my dad says, "can go back there any time he likes and take stuff off the sideroads."
Old Ritter never said a darn word. He just let that drop quick. The company was doing the
same thing.

WILLIAM McKEOWN (b. 1885)

I worked for Peter's Company back of Eagle Lake one winter. Bob Ball was foreman. It was
very big pine up there. I was on the dump on Irwin's pond, three miles from the camp. I've
been on that dump — go down with the lead team — at half past four in the morning. Jack
McEwen and I were making the dump. We had to dump logs where the ice wasn't good,
dump around and get the water freezing hard at night. We wouldn't get in until near nine
o'clock some nights. There were maybe fifteen teams. When the tail team got in it was near
nine o'clock at night, but they didn't need to go out till near ten the next day. The supper
would be all over when we'd get in, and there'd be some men in bed.

   We stayed with it and got the dump in real good shape, then he took us off and took us up
to the bush and put us loading. I was a greenhorn; I don't know how I did it, but I did. Talk
about work. It was big timber, and there'd been a lot of rain, and the logs were all frozen
together. All big pine, terrible hard to handle. We would have to put the decking line on
and pull them to get them down, put big pries in the end and put the decking line on. And
we'd have to get up and chop the full length of them logs, ice about that deep. I tell you we
chopped as hard as ever we could — just took long enough to eat. The reason he took us off
the dump, he couldn't get anybody else to load this stuff. Wouldn't do it. We did it. But, oh,
it was tough work. Those big logs were so hard to handle going onto the sleighs.

   The biggest log I ever knew of was sixty inches across, five feet, back in at Inholmes. Dick
Robinson took it out for the Parry Sound Lumber Co. Jack Parton skidded it. We had to
ross it. We had to ross the first two logs of pretty near every tree. Take a strip of bark off and
put a skid under it, and swing it this way then that way. Keep the slippery side down. Dick
Robinson's son Jack was driving a team, and he was so scared of hurting the horses —
babying them up so — that we had to pretty near grade the trails for him to draw on. One
day Dick came out and looked a while, then took the lines out of his hands. He just slashed
those horses up and down that trail to anywhere he could grab a log and hock it out. And it
poured rain. He had a white shirt on, I remember. He never stopped no matter whether
there was a trail or not. Drive 'em in over anything and hock the logs out. He never said
nothing, he was that mad that Jack wasn't doing very much. He put an end to that. "Don't
ross another log unless they're real big ones," he said. His son would have put him in the
hole.

ALBERT SCOTT (b. 1895)

I was born on the Nipissing Road, about four miles north of Seguin Falls. I first worked in
the bush when I was about twelve, cutting trails and peeling hemlock tanbark in the bush
behind the farm. My father owned the bark and he had sold it to the Ontario Hide and Skin
Company. Their tannery was down in Acton, and that's where it was shipped to.

   There were four men in a gang. One fellow was the chopper; he chopped the trees down.
Lots of them were two and three feet on the stump, and they were chopping them down. But
a lot of them were hollow-butted, just the bark, about four inches of rim, and the rest was all
hollow. The bigger they were, the thicker the bark. The first thing you do is take an axe and
ring [the tree] all around the roots, then reach up four feet and ring it again, and split it
down and take the butt sheet off. There is more bark in that butt sheet than there is in two or
three when you get up among the knots and limbs. It's heavier. Then you notch the tree and
fall it. There's generally three different directions a tree will go; it might be leaning a bit, but
if you know how to notch you can draw it off to the left or right of where it's leaning. You'd
always fell a couple of saplings underneath to keep it up so you could get underneath to
spud [the bark] off and pull it out. It didn't matter if you broke [the trunk] as long as you
had pieces [long enough] to make four feet.

   One fellow did the limbing. There was a hell of a lot of work about limbing. Them hemlock limbs, some of them touched the ground pretty near; they'd be eighteen and twenty

feet long at the bottom. Then you'd get into bushes where there'd be hemlock standing there forty feet and hardly a knot in it. One fellow did what was called ringing and splitting. He measured it off four feet with a two-foot axe handle — two lengths of the axe handle — and he chopped a ring all around, then split it so you could get the spud in to spud the bark off.

We shipped [the bark] from Seguin Falls in boxcars if we could get them, and if we couldn't we put it on flatcars, put stakes in them and wired them across at the top. We drew it to Seguin Falls on wagons in the summer and sleighs in the winter. We drew two cords on the wagon and four cords on sleighs. You turned the bark upside down when you were loading it. When you piled it in the bush you turned her barkside up, but when you loaded it off the pile onto a sleigh you turned it upside down because one [piece] fitted into the other better and you could get a lot more on the same rack. Taking it out of the pile always made it more bulky; it always drooped down on the ends after you piled it in the bush.

You put two [rows] on a flat rack. We used cedars twenty feet long and six or seven inches at the top end, and them went on the outside, on eight-foot bunks. They had stakes in them front and back. The rack just had a few boards in the centre, so when you put the bark on it had a slope to the centre and you didn't need any binding. But you had to bind it on a wagon, because you hit so many stones and one thing and another that would jar it, and it's awfully slippery. Cut an ironwood pole maybe three and one half inches at the butt and as thick as your wrist at the top end, and put a chain from the corner of the rack with a half hitch in it, and run the end of the pole in there. [The load] came up to a peak in the centre, so when you pulled down on the other end and made a half hitch on it, it tightened down on the bark. You'd only bind about five or six loads till the pole would get bent and you had to go out and cut a new one. There were lots of upsets with bark. When you get four cord of bark on a set of sleighs it's pretty high. When it's thrown on sleighs you'd think you had eight cords on, there's so much space between it. One time we upset a load into this creek that run from the far end of the limit. The creek was running so fast along there it never froze, and the bark went down under a bridge and jammed and backed it up right to where it upset. You take four cord of bark, there's quite a lot of pieces in it.

There generally would be two men loading the sleigh, the teamster and a fellow who stayed at the pile to help load. With the extra man to help you, you would load four cord in about half an hour. [At the railway siding] you'd drive up this tramway. If it was a wagon you had to stoop down to throw it in, because with four feet of bark on the wagon, the bark was above the top of the car door. Sleighs were a lot lower. We piled it in the car the way it came off the pile, the drooped ends [down]. They were paying $2.75 a cord piled in the car, but they didn't measure it eight feet by four, they paid for it by the ton. [A cord] was supposed to weigh 2,240 pounds. The cars have got their weights on the outside of them, and they weighed the car and whatever the net weight, that's what you got paid for.

MEL CAMERON (b. 1892)

I done near everything in the bush. I've rolled, I've sawed, I've chopped. I drove horses most of anything, for Parry Sound Lumber Company. I rolled in there all fall with Charlie Berger. Tom Bell from McKellar was foreman, Dad was the cook, George Knight was the bookkeeper and John Gadway was the walking boss. I rolled all fall, then loaded after the haul started. The foreman went around to his men and he figured who was right for loaders. He tried to get a right-handed and a left-handed sender. If they were both the same hand of hooksman, there'd be one of them backing up all the time. Those logs had to be kept square on the skids. You had to be alert and handy or the log was gone. Your top-loader would soon come down and kick you if you didn't keep your log square. That's loading with the decking line. Of course after they got the jammer it didn't matter which hand you were, right or left. All you needed were two bullrope men to steady the log when it went over the load. The top-loader would drop his canthook in wherever he wanted it, and they knew when they saw him drop his hook where to drop the log.

Loading logs on a sleigh with a jammer. — GEORGE KNIGHT

This jammer crew includes a top loader, two bullropers and two tailers-down.
— JIM LUDGATE

Ben French (r.) and his brother-in-law Joe Farley beside him, figure in several of Marshall Dobson's stories about life in Albert McCallum's Number One Camp in Hagerman Township. — MEL HEALEY

Norman Cameron (l.) working as a sandpiper. Sand dug from the hole on the right was piled on the fire to dry. — NORMAN CAMERON

40

MARSHALL DOBSON (b. 1893)

A skidding gang, there were two rollers, two trail-cutters and a teamster, that was the outfit. There's always be an older man. Art and Roy Macfie, Clint Andrews, Ben French and myself, that was our skidding gang my first year of driving team in a camp. I was seventeen that year, Art he'd have been sixteen, Clint was eighteen and your dad would have been nineteen. But boy, we skidded a lot of stuff. It was a company team. I took a kicker. She had won the day, and they asked me if I would try her, and I said yes. I took her to the bush and she never bothered very much. Oh, she'd take the odd spasm, but she'd quit and go on and work for a week, then take another bad spell. But if you whipped her you sure had a fight.

Ben French, we was an awful character. There'd be lots of fun with Ben, you never knew what he was going to do. He'd dress one foot with one sock and have three socks on the other foot, and wonder why one foot was cold. He'd be storming all day. Every morning he'd lay down on the floor, roll over a couple of times, then start for the bush. When he'd get pretty near there, you'd think there was a couple of dogs coming. He'd be howling and roaring and on the run. This day he was coming down the road behind the rest and he rolled halfway down the sand hill. Just laid down and started to roll. He was rolling down, and he jumped up and kicked and rolled again. McCallum says, "There's that crazy French coming." Joe Farley, his brother-in-law, they were loading, and this day Joe had a frozen toe. They were lifting the sleigh to put it square with the skidway, and Joe was on one side and Ben on the other. As soon as he got the sleigh up so's it was over Joe's foot he let go, and it went down on the sore foot. Farley just turned white; oh, it was a sore foot. Ben ran and took all the tobacco out of Joe's pockets, pipe and everything, and all day he would say, "Would you like a chew, Mr. Farley?" And he'd be taking one. He kept smoking Joe's pipe, and Joe had no pipe to smoke. So Ben froze a toe one morning, and at noon he took his boot and sock off to thaw it out by the fire. Oh, it was hurting and he was complaining to beat the band. Joe asked if he could take a look at it, and when he bent over he let fly all over Ben's sore toe with tobacco spit.

There were two yoke of oxen. Tommy Moore drove a quiet team and this Whitehead drove a pair of two-year-olds that were as wild as hawks. Roy Macfie, Ben French and another lad were loading, and when they came down Whitehead took and hooked them to a tree, fed them their dinner and went on down to his dinner. Ben came along and saw this wild team hooked to a tree, so he went over and broke a tag-alder and hollered, "Come here!" They just shifted and away went the balsam tree. Ben comes down to the dinnering place, and "Mr. Whitehead," he says, "your oxen's gone up the road. They got the tree with them." Whitehead dropped his dinner and away he went. He had to go to the camp and bring them back. Oh, he'd pull the darndest things. On the Shawanaga drive we slept in tents, eight men to a tent. There are rattlesnakes down there, and Joe Farley, he was very particular about snakes. He always shook his blankets before turning in. So Benny cut the tail off a muskrat, tied a long string to it, and after Joe went to bed he slipped it in by his feet. When all was quiet he said he felt a snake go across his blanket toward Joe, then he started to pull on the string. Joe landed on top of the man across the far side of the tent from him. He never caught on till next morning when he saw this muskrat tail and string lying in front of the tent. Of course Ben left it there for him to see. But Ben was a great man. As a rule he top-loaded, but when Joe was there Joe top-loaded and Ben sent. Joe didn't like Ben up on top because he just tramped a load. He'd get them on, but oh he was rough. If a log was a little too far in front, it didn't matter to Ben.

NORMAN CAMERON (b. 1894)

At McCallum's camp some loaded with the decking line and some with spike skids. The fellows that had to load with spike skids were kicking all the time, but if they gave them a decking line they didn't know how to load with it. It took training to get to be a good canthook man. Dave Simpson was a good top-loader. A good top-loader was a man who could get each log to the right place, get the chain back to the fellows down below real quick so they could get it around another log, get it coming up, and leave a place so the log would tumble in instead of rolling back down onto the skidway. You see the pup pulled out when the log went over the end of the chain. A good top-loader could balance his load, not put all the big logs on one side and all the little ones on that side. He'd have his logs so that when he put his last log on, the sleigh would rock, an eleven-foot bunk would rock.

McCallum tanked his roads. He had holes dug all down [Snakeskin] Gulley that filled with water. They'd fill the tank every little piece along the road and ice the sleigh tracks. They had half-round shoeing on the sleigh runners so they'd track easier. The half-round shoeing — it wasn't really half round — would be four inches wide and rounded from maybe an inch thick in the middle to nothing at the edges. The company that turned out those sleighs, the ones with soft steel on them for shoeing, they painted them blue. When they hit sand they'd stick you solid. The hard steel ones they painted red. They would make the horses run on a sand hill that the others would stick on. They put the red sleighs in one bunch and the blue in another, so if the red sleighs came first they had to put a little more sand on, and when the blue sleighs started to come they put a little snow on here and there. On some grades where it was rocky and they couldn't get a sand hole, they put beaver hay on. There'd be the "hay hill" and the "sand hill" and if they used manure from the barn there'd be the "shit hill."

I tended to a sand hill for Mac Harris when they logged this country back here. My brother and I were cutting and skidding in the fall, and when they started to sleigh haul they asked me if I'd tend this hill going down to the lake. I had a brush house built and burnt the sand. Mac Harris used to get the chickadees, the fellows that repair the road, he'd have them cutting wood for me and all I had to do was burn sand. It wasn't sand, it was clay, but that dry clay would stick a team dead. I saw a run once — my brother was on top of the load — he came down and the trace chains were too slack. One unhooked and away he went. The sleigh started to run on the horses and he had to let them go. There was a steep part, then a little slope, then a steep part down onto the lake. There was a turn where it went out onto the lake, and a big pine tree. The tongue dropped down, but he made it all the way to there, and him on top. The horses went that way and the sleighs went this way and hit the pine tree. The wrapper broke and logs went for fifty yards down through the bush — and he was in among 'em. He came out, not a scratch on him, and his team was standing there, not a scratch on them. And the whole load of logs just flattened out among the trees. Just a streak of luck.

At Mathews' camp on the French they had a tow hill. About 400 yards from the river they came to a hill. It wasn't over a hundred feet long, but it was steep. They had a winch at the top of the hill, a big stiump, flared, with a pin down through it that was frozen into the ground, and a cable around it. You pulled in to the bottom of the hill, unhooked off your sleigh and hooked a chain to your gooseneck, through the ring on the end of the tongue, to keep the tongue in the middle of the road. The chain hooked onto the cable on the winch. They kept the hill iced. There was a team of horses that stayed on the winch all the time because there was a steady string of teams drawing. They would wind those loads of logs right up the hill.

That winter, to save clearing a new log dump, they filled the Little French level with logs across the other side so they could draw across to where a dump had been cleared years before by somebody. They put snow on the logs in the river and went across and dumped on that side. When the flood came in the spring those logs were frozen in so bad the water ran right over them and they still sat there. They had to dynamite that all out to get them going. They blew half the logs to pieces.

You take a frosty morning, when you hitch onto one of those log sleighs and start down the road, it's just like hauling on sand. You'd hear the sleigh runners screeching on the frost, till the sun came up and a few sleighs went over the road. The first team down in the morning, you could tell when he left the skidway, hear him going down through the bush. Some of the old teamsters got wise to it, and they used to bore holes in the runners and put plugs in them. On cold nights they'd fill those holes with coal oil, and that coal oil would seep out onto the shoeing and away you'd go on the frost. You wouldn't hear those sleighs squeaking. Another fellow that had no coal on his shoeing, you'd hear his sleigh squeaking two miles away.

42

Timber grapple and storm boom chain bearing the Parry Sound Lumber Company mark.

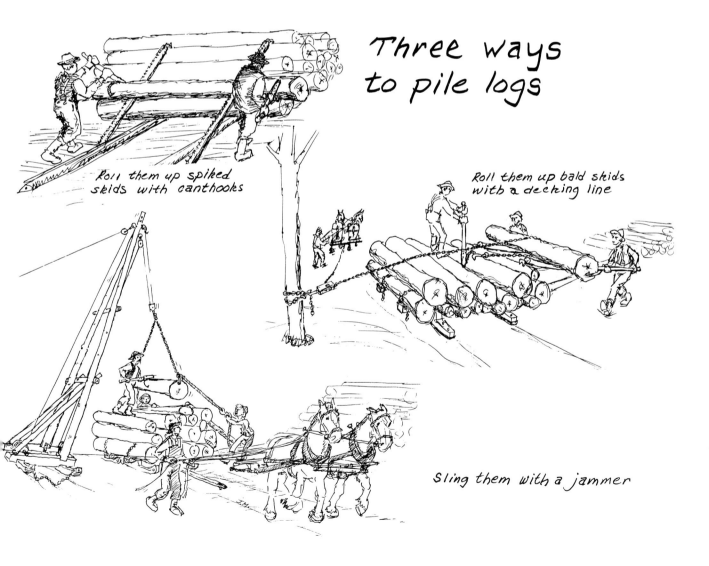

Three ways to pile logs

Roll them up spiked skids with canthooks

Roll them up bald skids with a decking line

Sling them with a jammer

The Tonawanda Lumber Company was the first to take timber out of here [Golden Valley]. They took all hewed square timber. Everything had to be twenty-one to forty-two feet long, all squared and a twenty-one-inch face on each side, or bigger. Just the great big stuff, just the choicest and perfectly sound. If it had a knot the size of your little finger it was culled and left in the bush. I can show you yet today some of them old square timber sticks all rotted down till there's just the top square laying on top of the leaves. The square timber was loaded at the French River and shipped to England. The sticks were made six inches longer at each end than the actual footage, and they were tapered with the axe for hitting the rocks and going through the dams. It wouldn't broom the end. The river-drivers weren't allowed to use caulked boots, they had to use rubbers or shoepacs so as not to mark the logs. I remember old Hughie Haines and Billy Kegley hewing timber back here when I was about eight years old. Billy Kegley used a twenty-pound broad axe with a handle just that long [indicates about twenty inches] — on his knees all the time. They had six men scoring the timber, and they hewed it. They squared in the bush till the snow came, then once they could sloop them they brought them out on the marsh and hewed on the ice. A sloop is just two runners with a bunk in the middle. Birch runners, with a crook in them. They were jackknife sloops, made with a loose bunk, all pins, no bolts. There was no ironwork in them at all. They were tapered, with a narrow, rounded nosepiece, and tongue in the centre. If one runner hit a tree the sloop would jackknife until it would go past.

EVERETT KIRTON (b. 1894)

I got my scaler's licence one summer, and Norman Cameron's, on Dobbs's Lake back of Golden Valley, was the first place I went to scale. Jack Sweet and I were at his camp next winter. Jack Sweet would never make a scaler, he didn't have enough education, but he was an excellent helper. I was the scaler, but I worked under Jack Sweet, I let him run it. He knew where we should scale first, and so on. Norman Cameron had about 12,000 logs, and the cutters' count, the haulers' count and the scalers' count were all within ten logs. I never had it that way since. Sometimes there'd be a terrific difference between the cutters' count and the skidders' count, or between the skidders' and scaler's counts. I think that was remarkable.

I fire-ranged for the Schroeder Mills and Timber Company around 1911, before they started operating. The timber companies hired the fire-rangers then. Tom McDougald from New Brunswick was the head fire-ranger over three of us. He generally stayed with me because that was centrally located. I ranged the south half of Wilson Township, and Billy Forsythe ranged the north half of Brown and stayed in a hunt camp. George Knight stayed in Miller Lawson's house in Ferrie Township. I made my headquarters at Island Lake or at Loring, where my home was. We travelled around and put out a fire if there was one. We carried an axe, a shovel and a pail. Sometimes we would go up Island Lake and portage over to Kelcey Lake. From there we'd go to John's Lake and into the Still River. We'd go down and see Billy Forsythe. And sometimes we would go over and see George Knight. We had a couple of fires. One was lightning and one was some tourists.

ROY COCHRAN (b. 1905)

Your best time for skidding was up until Christmastime, after that the snow got too deep. Unless it was fresh felled timber it was hard to get a hold of once it was buried. There were two rollers and, depending on the class of bush you were in, two or three trail-cutters, and the teamster. The teamster was the boss of the skidding gang. He looked his logs over and saw how much he had on this side and how much on that side, then selected a place for the skidway. He'd start the trail-cutters out to make one long trail both ways, then cut trails off of that one way and the other to get in where the logs were. If they were small logs he'd pull one or two out, leave them, swing around and go into another trail, get another log or two, come back for those two, and away he'd go to the skidway with four of them. Then when all the logs were skidded that were going there, he'd move over to the next one. Very often there would be logs in [the skidway site], so he'd pull them sideways to leave room. You'd cut a set of skids thirty or forty feet long — any hardwood except beech or birch, because they are slippery. Maple, ash, elm, hemlock; any of that kind of thing. You'd pick a place that was sloped downhill and put your head block in, a chunk of a log, then set your skids on it. Then you'd block them up so they wouldn't bend, and leave the tail ends right on the ground and sometimes buried in the ground, so that you pulled the logs up on the skids where the rollers would be able to handle them easy. Maybe you'd run into a thick bunch of timber, then you'd have to put in a second set of skids.

HENRY NORTH (b. 1903)   LEVI NORTH (b. 1898)

That was quite an outfit, Holt's. There was an awful big fire; somebody touched off some brush in the spring, figured they'd get a job next winter when they'd have to take [the timber] out. The pine was as black as coal. Everybody was black when they came back to camp at night,.

They had nine camps up there. I worked for a jobber named Whalen at Camp Eight. The stable there was for sixteen teams. There were 200 bunks in the bunkhouse. In Camp Nine 150 men sat down at the table all at once, and you could hear a pin drop. No talking at the table, just "pass the salt," that was it. Those were the rules. If they didn't listen the old cook would be right after them.

In 1920 I decked logs at Camp Nine with Dad's horses, loading on sleighs with jammers. I think I got $3.50 a day. A heavy team, they paid another twenty-five cents a day. There were five loading gangs and fifteen teams on that road, drawing five miles one way, and the train was also [hauling logs] then. Breakfast at a quarter to five. They loaded up three loads the night before. It looked like a midway, all lit up with torches. They took forty-five or fifty logs to the load. I timed it once and we were fifteen minutes putting forty-five logs on. Big logs, I could just lift one, but small logs they put the chain around three or four, and when they'd go up they'd just trip it and they'd pile in a heap. I've seen these big loads come down, the lines sloped pretty near straight down to the horses' heads. I think I drew the biggest log that came out of there. It had a fifty-two-inch top. This American, Frank Lucas, he was loading this log and he said, "You might as well roll it over there and leave it, there'll never be water enough in Farm Creek to take it out." We put that one and four little ones on and that filled the bunk. It was a marsh we were dumping on. I know some of the hunters that hunted in there and I asked them if they ever were on Farm Creek. "Oh yes, that's part of our hunting grounds." I said, "Did you ever see a big pine log in there?" "Yes, there is a big one there, now that you mention it." I said, "I guess you saw my log, then."

There were seven teams drawing at Camp Eight. I was there forty-three days and there were logs drawn before I was there and some after. I lost two trips. One was for too much on. He wanted to clean up the skidway, this gink from Orillia, Allan Stand, the top-loader. "Why don't you take them other three?" I said, "I'll never move it." They put them on anyway because they wanted to move their rigging. All they had to move was a decking line, a pair of skids and a snub. So they put the three on and I broke a tug. "Oh, we've got a snaking chain, we'll darn soon fix that." So they put the snaking chain on for a tug. I broke a hook on the other side, and I had to go to the camp. A team got stuck ahead of me one day and I lost a trip again. If you didn't make your trips they'd dock you.

ALEX GALIPEAU (b. circa 1908)

They told me I was a good top-loader. The top-loader was the one that engineered the building of the load. When you're hauling, the top-loader is boss. The foreman says, "I want as much logs on the dump a day as you can put there." Well, the teamster sometimes would say, "I've got pretty near enough there." Sometimes a top-loader is a little bullheaded and says, "I think you could take another couple more up here." And the teamster's got to take it because the orders from the boss are to put as much on the dump as he figures he can without any breaks or anything. Pretty hard to fool an old top-loader, he knows the roads, he knows what the horses can do.

Some top-loaders and teamsters would watch for a certain skidway where they could load the "brag load," as they call it, the best load of the winter. They'd ask the foreman to give them that branch line. It was mostly the teamsters, to show what they could draw, and

Guy Smith, holding the reins, drew this large load of logs into Lake of Two Rivers in Algon-quin Park. It was done for show, on a Sunday, hence the crowd of spectators.

There can be no doubt this big load of Schroeder Mills pine was actually drawn to the lake. It was built by top loader Bill Isaacs, a Micmac Indian, who is standing on top of it.
— JAMES LUDGATE

A brag load on the Pickerel River.
— GEORGE KNIGHT

Data accompanying this photo of a brag load says it was taken in the vicinity of Burks Falls and it was actually drawn to the dump. — PARRY SOUND PUBLIC LIBRARY

This gang went for count with their brag load, getting 175 on by choosing small logs. Jack Campbell is second from the right. — JACK CAMPBELL

the top-loaders, to show what loads they could build. Most of the brag loads are done on a Sunday near the end of the season.

Here is how you put on a load of logs. You have the man that tails the logs down, the top-loader, and two bullropers or hooksmen. With the jammer they're called bullropers and with the [decking] line they're hooksmen. A right- and a left-hand man; one would cut ahead and the other cut back. That's the loading gang. And three team of horses to a gang. We'll say the skidway has been shovelled out and broken, that means chop the little log off the front that holds the whole rollway. The jammer is all set and a team comes in and he unhooks. Every team pulls his own load up. They put skids with spiked ends made from horseshoes up on the sleigh bunks. The skidways are pretty nearly always level with the bunks. The top-loader stands on the sleigh. We'll say there are four or five small logs then a big one, and he'll say, "That big one." They'll put the pig's feet in that big one and he'll give [the teamster] the go. There might be four or five logs in front of it, and it will push them all up. Then he brings another bunch up, and he sees a nice face log, so he hooks it and puts it on the outside, and you put the binds on. The corner bind is a chain that goes through the end of the bunk, and it has a round hook and a grab hook. The log goes on the bunk and the chain goes over it. The round hook goes to the top of the log, then you grab it. There are four of those chains because there are four ends on the bunks. Put two in the corner binds and four or five in the centre, then loosen the back corner binds, find a vee-shaped log and jam it in. Top-loaders will tell you your front has to be narrower than your back so [no logs will slide] ahead. It can be tipped up at one end when you throw it in that crack, but the more load you put on the more it spreads so it will never slide.

When your bottom is down you start to build. You might make it five face logs high, a straight-up face, then throw the wrapper on it. If they wanted an average of sixty logs, well, five face on twelve-foot bunks and nine or ten logs on top of the wrapper to bind the load, and that's your load. If you leave enough slack when you put the grab hook on your wrapper, the wrapper will go down with the weight. On brag loads they used two wrappers.

In the fall they'd start to build up the log dump. They'd have a pole they'd drag with an old horse as soon as the ice started, bust holes here and there, and drag the snow with a pole to build up the ice. They'd send a young fellow out there with an old horse. Even if the horse sank, he'd done his hard work. There were three teams per gang, hauling. It depended on the size of the jobber. If he had 40,000 to 50,000 logs on skidways, he might have to use four gangs, that's twelve teams of horses for the haul. It would depend on his length of hauling, too. We'll say he had an average of five miles, that's the main road, to the lake. That's a ten-mile trip. Ten miles, that would be two trips a day for a while. Then once they started to come in a little early, he might make it two trips and a half. Make your two trips and load for the night. Then it would be two trips and one halfway to gain again. Then three trips. We'll put it a man with 40,000 logs. That would be three gangs, nine teams of horses hauling down that road. It's seldom there's a mile of road you could call straight. There are lots of crooks and turnouts. There might be four or five turnouts in a mile of road, and you'd hear the man with the load shouting. He would shout every now and then as he was coming down the road, "Turn out," or "Clear the road, here comes Johnny," or whistling or singing. And the other team coming up from the dump would hear him and turn in and wait until he was by. The guy with the load of logs can't stop, so the other lad's got to shoot into the bush.

GUY SMITH (b. 1885)
There were four men in a bark-peeling gang. With the run of hemlock they had at that time, four men were supposed to peel forty trees a day. There was the chopper, who chopped the trees down. I don't know why they always chopped them down. It's a big job, done the right way. I often wanted to do it. The chopper and limber have a saw and go together to saw the trees down. The chopper takes the butt ring off first, tight to the ground and up four feet, cuts it around, splits it down, spuds it off, then chops the tree down. That's his job.

Then there's the limber or trimmer, the fellow that does the limbing. And another fellow does the ringing and splitting. Every four feet of the log he cuts a ring all the way around and splits the bark so the spudder can spud it off. It's a dirty job, spudding, a mean job, because they don't get the knots trimmed off tight to the tree, and when you come along with your spud and take the bark off, the knots stick up an inch or two. They're sharp as an axe, and you're skinned right up to the shoulders on them. With thick-barked hemlock you get long knots, and the spudder has to get all that bark out from under the tree. When you fall a tree you try to lay it up, put it over top of another one so the ringer-and-splitter can get right under it.

The spudder stands the bark up on both sides of the tree, with the bark up, to dry it. Next day you'll quit a little early and pile it up; what you peel today you pile tomorrow night. You [cross] pile it in a pile four feet square.

I've spudded and I've ringed and split, and I've limbed. That's a hard job, limbing, and it's a big job chopping them trees down. The chopper used a double-bladed axe. He'd have a good sharp side for chopping the tree down and the other side a little bit thicker for ringing and splitting the butt ring. Tanbark will dull your axe more than timber; you don't use a sharp axe on hemlock bark. But the fellows that are doing the limbing and ringing and splitting use single-bladed axes not ground down so thin, because they don't have any heavy chopping to do. Hemlock knots will break a chunk right out of a sharp axe, spoil it altogether. The limber has to have his axe sharp, but he has to have it thick or those knots will break it. It's very seldom the chopper's axe breaks, for he's never in the knots. It's the fellow that does the limbing that busts his axe.

I think the farmers were the best bushmen. Not all of them, but most of them. You could depend on pretty near every one of them being a good man — all used to driving team, and rolling and loading, making logs. A fellow told me: you take a man that you can hear coming half a mile before he gets to the camp, yelling and singing and just drunk enough to make the camp, he's the best man you'll ever get. He's all lumberjack. That's true too.

BILL SCOTT (b. 1887)
One camp I was in, we took out 1,900 cords of bark and the logs were just left in the bush. We just cut through them wherever we wanted a trail to get the bark out. We took trees down to the size of stovepipe. On deeded property [being cleared for agriculture] they were logged up, put in great piles and burnt.

WALTER SCOTT (b. 1893)
There used to be three men in a [cutting] gang, two men sawing and the chopper. There was a fellow named George Strathearn who worked for people called Clark and Horne at Blackstone Lake, and he had a glass eye. He was a chopper. If anyone came along when he was working he'd split the measuring pole on the end, take this glass eye out, stick it in there and shove the pole up the tree to see if there was any punk in the knots before he cut it down.

A lot of places they'd cut the biggest part of the logs before they started to skid, but it's not a good idea to cut them all. I used to cut a few before I'd start to skid, but not too many. One winter Jack Campbell logged up at the Magnetawan, they left it to take the horses across on the ice. But it would have been better to build a barn across the river and swim them over in the fall, then skid as they went along. It got so they'd just skid ten logs a day, there was so much snow to shovel. You can fall trees anytime, but you can't skid anytime. October was the best month for skidding as a rule, as long as it was wet. If it was a dry fall, logs were pretty hard to drag.

Every [skidway-building] gang would have three trail-cutters, maybe four. Trail-cutting was about the worst job. A lot of people didn't want to cut trails, they wanted to saw or roll or drive a team. The teamster would tell them where to cut. A good teamster would skid from both sides of the skidway; he never turned around and went back the same trail. He'd bring a log from that way then go out and bring in a log from this way. Turning around to go back delayed you because you had to wait until the roller took the log out of the crosshaul before you could go back. I've seen skidways with 400 logs on them, but most of them were around 150 or 200 logs, because there's too much tailing to them. A fly skidway is

50

made on a steep hill. You put small skids in at the bottom, a head block and skids about twenty feet long. Then as you go back up the hill you keep adding skids. If you put long skids all the way down the hill, you have trouble getting the logs rolled down; they go all over. When a log is on the ground you can catch it with the canthook and stop it any place. The rollers would have to not let a log get going too fast; catch it once in a while to hold it. I remember one time a fellow was working on a fly skidway and he let a log get going to beat the deuce before he caught it. He grabbed it from the back, reached out and grabbed it with his canthook, I'm sure he went twenty-five feet down the hill. I thought he'd kill himself in among the stumps. You want to be in front of a log when you're going to catch it, because if it's going any speed at all it'll throw you clean over top of the log. It's liable to kill you, the log will roll over top of you. There's generally one roller who stays ahead of the log and the other one behind. Rolling down a hill like that, always one fellow stays in front so he can catch the log whenever it needs turning. The big end is always going to gain on the small end. If you had to roll fifty feet it would get clean crosswise of the skids if you didn't cut it before you got down there.

I never piled fly skidways up too much right at the start, because if you piled them too high, when you broke the skidway you'd maybe get two loads going out over the sleighs. I have seen skidways where people put them away up high in front, but that's an awful mistake, because once they break, any that comes loose they're going to roll. They kept two fellows breaking down [while loading sleighs]. They always kept four or five logs ahead and let the rest come down against them.

Guy Smith and I were skidding off a mountain down at Whitney one time, and there was no way to get a road off it. The only way to get the logs off that mountain was shoot them off. I'd skid them to the top and unhook, and a couple of men would give them a roll and away they'd go. Then Guy would take them out to the skidway.

They figured that teams on the draw road should make thirty miles a day, that would be fifteen miles loaded and fifteen miles without any load. It was a good job. All you had to do was deck your load, and you were sitting down the rest of the time. The teamster helped dump, but there'd be two or three men there. They liked to get their load off as quick as they could and get back on the road. They used to get off and walk behind the load — tie the lines to the binder and get off and walk, let the horses go themselves. It wasn't too hard a job, but it was a long day. When I was working for Bill McKelvey, we used to feed our horses at half past three in the morning, and lots of times there'd be teams coming in at seven o'clock at night. But there'd be fourteen or fifteen teams hauling, and the fellows that went out real early, they'd be in about half past three or four o'clock.

The teamsters had a long haul, a three-trip [per day] haul, and I was driving the tow team. I had to be out on the road for the first team to come along in the morning, and I had to stay until the last team was over the hill before I could go home. I was just a young lad, and I got fifty cents a day, but the best men in there were only getting $18 a month. They used to pay $2.25 or $2.50 a day for a team and man. A lot of them used to go to the camp in the wintertime because they'd get their horses fed and they'd be in good shape when they came home in the spring.

A load of this size slipped easily over an ice road. — GEORGE KNIGHT

THE ROADS

WALTER SCOTT (b. 1893)

A giper had an axe and shovel and went along the road. If the sleighs had started to dig out a hole he'd cut a bunch of little poles, put them across, fill them with snow, pack them down and let them freeze overnight. Once they were frozen they'd never move. He'd run the axe handle through the shovel handle and carry it over his shoulder. He might have a mile of road. When they were drawing six or seven miles there'd maybe be seven or eight men fixing the road all the time. Chickadees, they mostly went along throwing the horse manure off the road. They'd have an axe and shovel both, because the road might get bad someplace and they'd have to help the other fellow fix it.

I always tried to have ice roads, if it was possible at all to get water. When you have good ice roads you can draw much bigger loads and it's no labour for your horses. You'd see them going on the level and their whiffletrees would be playing up and down. Steering the load was just about all the horses were doing.

When I went to the bush in the fall the first thing I did was cut all the roads and grade them down, levelled them right off. Roads that are cut in the winter when there's three feet of snow on the ground, you can't level the knolls off, you can't do nothing. I've seen [a jobber] with a million and a half feet on skids at Christmastime and the only roads he had cut was what two men done in part of a day because he didn't have any other job for them just then. He had to shovel that road through. He'd have ten men shovelling snow and six men chopping stuff out of the road. The six men would have cut three times as much road in the fall.

I'd start about the second snowstorm and run the snowplow through the roads and get the frost in them. Just put the vee-plow to them enough to knock the snow off, and let them freeze good and hard in the bottom. If we had swamps, sometimes we'd have to use snowshoes and tramp them to let the frost in. I've seen swamps you couldn't put a horse in, it would go right down out of sight. One time I was working for Ludgate and Thompson and we had a lot of tanbark to draw to Neibergall's siding. There was a great big swamp, and we started to break it out with two teams and a couple of decking lines, and four or five men to tramp the holes. The first team got through all right, but the second team got down so bad in the bog that we had to pull them out with the decking line — put the decking line around their collars and pull them out.

[To make a vee-plow] we got a couple of good big pine logs and hewed them down till they were about four inches thick. Then we took and bored two big auger holes in the front of them and put a chain through. Back about four feet from the tail end, we bored a big hole in each side and put a pole in to keep them from going together. But that was only for knocking the snow back and getting it to freeze. After that we had a patent plow that would cut the runner tracks out. You had to do a little tanking first, put some water on before you used the patent plow, or they'd get down and cut into the roots. It made a good job of the road because it cut the runner tracks out and the sleigh couldn't slew; it had to stay in those trenches. At night, before you'd start to tank, you'd go over the road with the patent plow, clean the horse manure off it and groove it out. When you were first breaking the road you needed two teams on the patent plow, but once you got lots of ice on it one team could handle it. It was on a pair of sleighs — the runners were pretty light because it pretty near carried itself. When it came [from the factory] the big steel wings that screw down would be off the sleigh, but the rest of the plow would be all together. All you'd have to do was slip the wings on and it was ready to go. It had a vee-plow running in front to push the snow out of the tracks. On the outside, over the runner track, was a steel flange, and it had a wheel with a heavy worm that pulled the blades up and down. It was just like the rig to brake a [railway] car. There were two men on it, the teamster and the fellow that was handling the screwing down, or whatever was necessary. They really were the only thing for keeping the roads in shape. If you didn't have a patent plow you had a lot of giping to do, because the sleighs would take a shot here and there on the road. Once you got the road grooved the sleighs would stay in the same track, they wouldn't slide around. But they cost an awful lot of money.

I'd have what you call a buck beaver [in charge of] cutting the roads. I'd go through and blaze the roads all out, and where the skidways were going to be, and they'd cut it out. I'd do all that in the fall. Some would just skid to the blazes, and the odd fellow didn't blaze the road at all. Jack Macklaim was working for this fellow, and Macklaim said, "That skidway's done where I am. Where will I put in the next one?" The boss said, "You know where that old turned-up root is beside the horse trail going out? Just put it where that old root is." Macklaim didn't know whether to face the skidway north or south, and when they came to cut the road the skidway was sideways to the road. You can't skid logs when you don't know where the road is going.

ALEX GALIPEAU (b. circa 1908)

Tanking roads is a nice job. It's cold, but you're your own boss. There are just two in the crew, you and the teamster. Every night it's cold you go out and build the road up. I was what they called the captain or conductor. We'd leave the barn around half past nine or ten and be back in the morning about half past three or four. We had torches with coal oil in them and a big gob of wick, a metal thing with no glass. There's another name for it, a "bitch."

That's the main road to the dump I'm talking about. It was very seldom you iced the sideroads. Let's say there was a slack off of cold weather, and he'd call us out in the afternoon to dredge the road with the patent plow. It's for when the road gets too high on the outside and nothing in the centre and those half-round runners would run off. They call that runner-bound. It's like two shovels that stick down and cut [grooves] for the runners. It's adjustable, with pegs. There were screws to lower the wings to cut into the ice. You'd be out there maybe three o'clock in the afternoon to do dredging, soon as the last trip was down. You can't do that in the dark, you have to see what you're doing.

So after you get this dredging done you put that sleigh aside and go and get your tank and fill it. You have waterholes to fill your tank. These tankholes most of the time would be in a swamp or little creek along the road. They'd blast a hole in the swamp. You'd have quite a few holes like that, and sometimes on the lake too. I've got wet a few times falling in the hole. It's cold at night and moccasins are the only way to keep warm. It's wet and first thing you know you fall in. There's a ladder you hook on the side, two poles with half-round hoops on which a barrel is drug up and down, a barrel with a pulley. In the bottom of the barrel there's a lid, and when it drops in the water it tips up and the barrel sinks in. Soon as they start to pull with the pulley the lid will close. When there is about three-quarters of the barrel past the hole in the tank, it hits two crooks on the side so it dumps. The horses back up and the barrel slides back in the hole. The teamster would look after his horses and I would say, "Ahead" and "Back." The tank would hold around forty or fifty barrels. And I would make damn sure my plugs were in solid. There were four plugs straight down and two plugs on the sides with tin deflectors — just an old shovel with a hinge and a chain you could drop or [raise] to throw water to the side. It would hit there and freeze. There were plugs at the tail end of long sticks with a crossbar through them. There's about a three-inch hole and the plug is kind of pointed. If I only pull it so far out it will squirt just a little bit around the sides. If I want to plug it tight, I'd take my axe and hit on top of it. If it is stuck, frozen, I bang it with my axe, roll it with the crosspiece, and it will come up.

After we had the dredger on we'd pretty well open four holes on the straight go, the water going straight down. As soon as you have your tank full and your holes open, there's hardly any work, it's the horses that do all the work. Your hardest job is to keep warm. You're up on the tank, opening holes and closing them, and watching behind. You have a torch on top to see your work.

If the road was nice and you only had one trip right through, maybe you'd be back in around two o'clock. The cook would always have something for us when we came in. We'd take a lunch with us too, stop on the road and make a fire and eat. The last part of February is bad for the main road. In February there's a thaw, the sun is hot and the snow melts on the one side of the road. You've got to build that up, that's where your work is. Sometimes you work all night to build it up with ice.

54

Filling a road-icing tank. — PARRY SOUND PUBLIC LIBRARY

Road-icing tank. The horses could be hitched to either end. — GEORGE KNIGHT

Starting down a sandhill. — GEORGE KNIGHT

MAN AND BEAST
TEAMSTERS, BLACKSMITHS,
HORSES AND OXEN

NORMAN CAMERON (b. 1894)

There used to be farms at the Miller Lawson place on the North Road between Maple Island and Golden Valley. They quit using that road for public travel about 1912. There was Miller Lawson, a fellow by the name of Gould, and Jack Dawson. Dawson cleared a big farm, but it was all stone and rocks. He used to ranch the Ontario Lumber Company horses on it. He had a fenced-in pound for putting the horses in until the lumber company came to get them. They had posts sunk in the field with logs across the top and brush over that for shade for the horses, 'cause in the summer the horseflies would eat them up. If they got in the shade the flies didn't bother them. One bunch of horses took a disease in their ankles, and they chewed their own feet off. I saw a nice black stallion put his foot up on the manger and eat it off with his teeth. Had to shoot him. Lost about half them.

JACK CHISHOLM (b. 1896)

The winter I was up at Deer Lake, after the sleigh haul started, I started to drive team. I had the biggest team in camp. They weighed 3800 just with the halters and collars on. And I took out the prize load. They had eleven-foot eleven-inch bunks and 110 logs on. There was $10 for putting it on the ice road. Every team had their try, and I put it on the ice road for them. One of my horses threw two hind shoes when he hit the ice. He just drove his toes in and, well, one shoe went clean back past the load and whistled past Burley Harris's head. I guess I was a hundred yards down the ice road before I managed to get the load stopped at a turnout. Burley said, "Are you not going to take it to the dump?" I says, "Not with two shoes off. Morgan's got two shoes off. Not on that ice road!" So one of the other fellows says, "I'll take it down," and he gave me his sleighs to take to camp. Old Charlie Croswell was blacksmithing. I took the horses to the shop and said, "Have you got a couple of shoes to fit Morgan?" Charlie was a good blacksmith, but he only had one leg and couldn't put shoes on. He caulked them and fit them, and I put the two shoes on. I was only a kid, too.

The biggest sensation on the sleigh haul would be a sand hill. I got a run on a sand hill one time. It was snowing that day and I guess the sandpiper, the man who puts the sand on the hill, he though he had enough on, but the snow was falling just a little heavier than he figured. And I got a run. There's always, as a rule, a bend at the bottom where you swing down into the ravine. I jumped. Instead of going down the ravine the team went straight on. They went straddle of a birch tree, and that's the only thing that saved the horses. We got the horses out, but we had no sleighs left. They were all smashed to pieces. We went back to work next morning on another set of sleighs.

JIM McINTOSH (b. 1896)

There was always somebody who stayed in the camp all summer looking after the horses. They kept the horses back in there, run all through the bush. Lots of hay on the draw roads because all the sand hills, they put the manure on them to hold [sleighs] from going down too fast. I've seen hay three feet high on those hills — timothy and red clover. There'd always be five acres cleared for a campground and there'd be hay growing there too. Those horses, they'd go all over the country, but once or twice a week they'd all land out where the old fellow was staying in the camp, for their salt. They used to go through the bush, look after them, see they were all right. They took their shoes off in the spring. I stayed back there four summers, and I sharpened up all those shoes and put new toe caulks on them. I'd have a couple of hundred shoes ready for fall.

Yes, I blacksmithed in the bush. You've got all the hooks to make, you've got twelve teams of horses to keep shoes on, you've got skidding tongs to make, you've got axes to hang. There's no one blacksmith can blacksmith for 122 men. You've got to have a handyman to put those stocks in, make canthook stocks, go out to the bush to get ironwood. You'd fall a tree and split it and hew it all down. Those axe handles there, I can lay them down against something and jump on them. You can't break them. I want a piece of ironwood five and one half inches [in diameter]. That goes into four pieces, then you take the heart out and throw it away and you've got all snow-white wood. Your handle should be made next to the sap, the outside of the tree. Now look down that, see the straight grain. A saw couldn't saw it any straighter. That's beautiful. I just shape it out with the axe, then I've got a drawknife and a rasp.

You had to take in tong steel and canthook steel. Tong steel is a mild steel and canthook steel has got to be so it won't bend when it lifts the logs. There's got to be a give to tongs. The horseshoes came already made, but you had to put the toe and heel caulks on. After Christmas you had to shoe the horses sharp so they wouldn't slip on the ice roads. But you couldn't have sharp shoes on when they were skidding. If they over-reached turning around in some narrow place they'd cut a foot. You didn't want a caulk in the bush at all, only have a shoe to save the hoof from breaking. But when it came to blue-ice roads the horses would have to be shod sharp. The inside heel caulk, instead of being crosswise of the shoe, was turned in so that if he hit the other leg turning around it would slide down his leg and not cut it. A lot of horses were terrible for over-reaching. They'd put the hind foot on top of the front one. If they had a horse doing that, instead of putting the toe caulk on the front of the shoe, I'd put it on the inside, so when the hind foot hit the first one it couldn't catch a heel caulk.

The most stinkin' thing in the world is that thrush in [a horse's] foot. It would sicken a dog, the stink. It was from people not cleaning the horse's stable, leaving the horse standing too long on the manure, not keeping the right kind of bedding. They should use the chaff out of the hay. Every once in a while, before you put in fresh hay, you scoop the bottom out of the manger and throw it under the horse.

They had all kinds of dope for horses, like sipadilly and "high-powered pressure" and powders of all kinds. I had a recipe an old teamster gave me. Long Dan McKinnon from Uphill east of Orillia gave it to me — drove team all his life. There were seventeen different things in it, asafidity [asafiteda?] and all this stuff, all ground up into a powder. Give 'em a poke of it every night in their oats. Once in a while they'd give a horse too much. He'd lay down and kick and roll until he made himself better. I never saw a horse die from it. A lot of them fed dynamite, but I never had any use for that. It made them stand a hard day's work, that's all. Unless you abused a team they would put on beef every day. I've had my team at Christmastime, four inches on 'em, just glittering.

You fed three times a day, the first one at four o'clock in the morning on the sleigh haul. They'd get four gallons of oats a day, a little better than a gallon a feed. A hundred-pound bale of hay would do a team of horses one week if you gave them four gallons of oats a day. Every Saturday the stable boss brought a big kettle of boiled oats and you'd put blackstrap on that, a pint for each horse, in his box on top of the boiled oats. Give it to 'em nice and warm. The horses used to really enjoy that. It kept their bowels in good order.

NELSON CLELLAND (b. 1899)

I saw the prettiest piece of drawing I ever saw in my life up at Ludgate's. They had a hill that came up off a little lake. It was quite a grade, but they'd get a run at it coming off the lake. You used to see them hit there on the gallop. The sleighs run pretty easy on the ice, and the speed and weight took them over. Oh, hell, a team would never draw a load [of logs] up that just walking. Murray Chisholm got up pretty near the top and he couldn't make it. So he held it, slid down off his load, snapped the doubletrees off the gooseneck and let the load go. She went back and curled down into the bush. Well, Archie Bear came along with his team, and Murray and him hooked onto the load. Murray was on the pole and Archie had a chain on it. Them teams got right down till their bellies were almost touching the ground, and at last it started to move — you could just barely see it moving. Once they got it going

58

they took it out of there and up the hill. I never saw two teams get down and hang like them two together. They just squatted, set down and hung on till that thing started moving. And took it out. They just got right down and hung! I never saw anything prettier in my life.

We were busy when the teams were there, but then they had to go away back so far. After we got the river filled up we'd be getting away out on the ice. We'd get one of those big pine with five or six hundred feet in it, get it rolled away out to where there was quite a drop, dump it over and run to beat hell. It would go down and hit the one below and, with the weight, sometimes you'd see twenty or thirty feet of that skidway break through the ice away down underneath.

We had a little brush shack up in the edge of the bush and we cut an old pine stub down for firewood. We'd go in there and take our canthooks along. When a teamster would come along his hands would be cold. We'd have a hook sitting beside the fire, pretty near red hot. We'd take that out and the teamster would pull off his mitts, grab that and say, "Oh, that's nice!" There was one teamster, he'd go out with an old smock on, a pair of overalls, and about $50 worth of spread rings on his horses. He thought a whole lot more of his horses than himself.

GUY SMITH (b. 1885)

I don't think a horse would last much more than ten years in the bush. I drove a team for my father-in-law for six years. Nobody else ever put a line on them, only me, for six winters, skidding and hauling. One got his rump drove in. They'd peeled some birch in the spring and there was a bad sand hill going down onto the lake. There was a big birch right in the middle of the load and it went out of the load and hit this horse. The kerf happened to be pointing that way, and it made a hole in him between his tail and the top of the rump. We were down the hill, so I went to the dump then to the camp with him. They had to shoot him. When horses got too old for logging they used to sell them. I've seen them sell a horse for $10, just to give him a home. If you wanted a horse to draw a bit of wood and monkey around, you'd get a horse like that.

I was a good teamster. I was reckoned among the best teamsters in the country. The teamsters don't get paid any more than the fellows that are rolling — and they have nothing to do, only roll logs. The teamsters are working nights sometimes until it's time to go to bed. It was just something to blow about, I guess. You gotta be a teamster to be a lumberjack.

W.T. LUNDY (b. circa 1910)

I drowned a team at Dellandrea's camp at Loring. It happened when we were dragging the lake to dump the logs on. The old fellow sent Bob Ewins and me out with a twenty-foot stick, about six by six square, with one team on each end of it, and we were dragging it up and down the lake. And I drowned mine. He got his team unhooked off it, but mine, I didn't get them unhooked. One horse got over top of the other, got his feet over top of the other horse. What happened, when the ice broke it didn't just break and let the horses go straight down. A great big piece of it broke and the end the team was on tipped. It drove the one horse under the edge of the ice, and the other horse's feet went over top of him. We yelled and the old handyman up at the camp heard us and came down, but before he got to me the horse that was underneath was dead. And before we got the other one out he was drowned too. It took two decking lines to reach out to where they were. They wouldn't take a chance on going out.

So when the old fellow came home at night — he was out buying another team — I had the harness alongside the big box stove, hanging up on some sticks getting dry. He came in and he says to me, "Well," he says, "I bought a team better than that team you're driving now." He says, "You won't be able to outdraw these." "Well," I said, "It won't be hard to outdraw the pair I've got." And I pointed at the harness. Boys, it just took the breath from him. I had told him. I'd tied ropes to the horses' halters and put them up on their hames. He'd said to me, "What's that for?" I'd said, "That's so when the team goes to drown I'll be able to get ahold of them." That was at noon and I had them drowned at four o'clock.

GORDON WHITMELL (b. 1899)

McCallum, you'd think he was a hard-hearted man, but one thing he said: "Men and horses gotta be fed." He wouldn't keep anyone who wasn't a good cook. Oscar Clapperton of Loring was cooking that winter, and boy was he good. And you'd think McCallum didn't care about horses at all, but he liked to see a team looked after. He wanted me to go back. Dad was selling potatoes in Parry Sound and we stayed in Sheridan's boarding house that night, and by god McCallum came in and jumped Dad right away. "Who learned that young bugger to drive horses. It wasn't you. You never brought a team out of the camp in as good a shape as that team went out this spring. They were the best team in the three camps!" He said, "If he'll come back to the camp I'll give him the best team I own." But he was a man to talk. If a horse got a little cut up, he'd say, "Oh, it's a long way from his heart." He'd talk that way, but he did like to see horses looked after. Always lots of feed.

ARNOLD McDONALD (b. 1907)

It pretty near always spoils a team when they get a run [on a sand hill]. When they come to it again they want to hold back. It's not hard to teach horses a new trick, but it's hard to break them off of something they know. You take a team that's been in a runaway, they're pretty near always in trouble. Your dad had that runaway team. I had a runaway team here, and you had to watch them very close or they were gone. It's hard to break them off that. In fact I never saw a team that was broke off it. Tudhope and Ludgate's had a team, a pair of Clydes — Hogy and Ralgy, I can still remember their names. You talk about a nice team to draw. They'd just squat and lift. A good draw team don't jump, they squat and lift and hang in there. In the bush, with trees falling and everything, they never bothered. At dinnertime you just put their oats and hay out and never tied them. But put them over a pole, onto a sleigh or wagon, then look out! They ran away one time or another with every man that ever drove them. That's where they got their run. Yet you could fall timber all around them skidding and they'd never think of running away. On the river drive we used to portage the boat from Harris's down to Shorts Bay, quite a big gas pointer, and we had a long sloop. We always had them hooked next to the boat, then the chain out for the other [team]. If there was a sharp corner it would cut them off, and they'd just squat and lift that theirselves. The last man that drove that team, I guess, was Murray Chisholm. He was a teamster. He had the lines in his hands and knew what he was doing. A lot of them, it would have been better to have the lines tied around a bag of shit — it would hold them back a little. These fellows that shove on the lines, it's all right in the cold weather when they're froze, I guess.

BERNARD MOULTON (b. 1894)

My father, Robert Moulton, came from Owen Sound to work for a harness-maker, Clark was his name, here in Parry Sound. He learned his trade at twenty-two. He was here maybe four years, and Clark sent him to Dunchurch in about 1894. It was all logging up there then. In about two years he bought Clark out up there and he spent the rest of his life there.

He sewed the harness all by hand. He had a kind of rig that you tramped down and it put the tug in there tight. He had awls to make the holes in the leather and made his own waxed ends, thread. You'd have about six or seven threads and roll and wax them, black harness wax. He had a kind of snap on the bench, and he'd put his cord in and bring it back the length he wanted it maybe a dozen times, then cut the ends and roll and wax her. He'd use two harness needles, one in each way and keep crossing. He had a tool, a rig with a wheel on it, that went along and marked where to punch your tug with an awl. He had all kinds of knives and punches. One knife was on a kind of gauge, and he'd bring that up the full length of the side of the leather to cut it.

[His main trade was] heavy harness for camps. Tudhope and Ludgate was one, and Turner Lumber Company — their camps were at Maple Island. Schroeder's and the tannery company used to send their harness to Dad too. I can remember Bob Burns bringing harness up from Grassy Bay Camp for my dad to fix. I remember he worked until twelve o'clock at night on them. I saw a hundred set of harness come in for overhaul in the spring, mostly tugs and backbands and belly bands and britching, that was the main part, worn

A tow team helps pull a load of logs up a hill. The fire and horse blankets suggest the tow team and teamster have idle periods between sleighs.
— JAMES DOBBS

Towing up a steep pitch with a donkey engine.

and some broken. Tugs would be pulled out at the hames and you'd have to ply them over again.

A new set of harness was worth $25 then. He got his leather in hides, in rolls, and he sent out for snaps and buckles and stuff like that. Tugs he used to make about four ply. Hame straps, bridles and halters would just be one ply. Making collars, you had to stuff them with deer hair or straw. He made the collar, then had this iron and punched that stuff down the top and made it smooth. His business went out with the [lumber camp] horses. The horses on the farm didn't amount to nothing. If they broke something on the farm, they'd tie it up with haywire rather than bring it in and get it fixed.

MARSHALL DOBSON (b. 1892)

I've been looking horses in the arse since I was a young lad. I most always drove team in the bush. I worked at old Number One Camp a whole winter. I tanked. We would generally go out in the wilds of night.

There used to be around sixty to a gang in Number One, that was cooks and everything. They had maybe sixty or seventy pigs running around. There was this boar we called Old Handspike. He had big long tusks, and one day he gave one of the bulls a bad gash. Old Mac decided to fix him. We lassoed him and took him into the shop, and Billy Craig sawed off his tusks with a hacksaw. Poor Handspike left home. When we let him out he took to the bush immediately. It was March and he never came home for three weeks. The next time a sow came around I had to go out in the bush and hunt him up. He was away out along Crane Lake, rooting ferns. Right in the snow.

Then they had two yoke of oxen and three or four cows. They had plenty of milk for the gang, winter and summer. you'd come along and think it was a farmyard. There were two big horse stables, a hen pen, a pig pen, and a lean-to on the one barn. And there was an office, a harness shop and McCallum's house. Oh, it was a real little town. The two teams of oxen snowplowed the roads. They'd go through mud where a team couldn't. They'd be right to their bellies and keep on going, take the plow with them. They've got a lot of power and never cut themselves. McCallum bought one team from the tannery company, 3200 pounds, and they'd draw any team in camp. If there was footing for them, they'd fetch it. They were full of bits of haywire from eating baled hay. When they butchered one of them, old Sam, they dulled the knife on a piece sticking out through his hide.

WALTER SCOTT (b. 1893)

Tom Canning was in the blacksmith shop when I first came to McKellar. I used to get my horses shod there all the time. He was a fine old fellow. Tom's father had blacksmithed in the same shop. Tom opened that blacksmith shop door at seven o'clock in the morning, six days a week. I'd often be there at six; I'd be in a hurry to get my horses shod and I'd want to be the first there. I've seen four teams in the yard waiting for Canning to come. He would shoe a team all around in about an hour and a half. Dave Johnston down here [Parry Sound] could shoe a team in less than an hour. You never seen such a man to get around. With the horses away over here in the corner, he'd make about two jumps from the horses to the anvil. I went down to town once and I said, "Any chance of getting a team shod before dinner, Dave?" There were no other horses in the shop. He pulled out his watch and looked at it. It was eleven o'clock. "Yes," he said. "Run them in. I'll shoe them." I came back at twelve o'clock and the team was shod all around.

Horses are like people. Some are a little bit stupider than others, but lots are pretty wise. A fellow was telling me a teamster was sick and the fellow had to take his team out the next morning. He said to the foreman, "I don't know anything about the sleighs, what sleighs they'll be." The foreman says, "You can likely leave that to the horses." So when he went out to the sleighs the team just stepped over this tongue, ready for him to hitch on. He said, "I hitched them onto that sleigh. I figured the horses knew more about it than I did." And he had the right sleigh, too. If it was convenient they used to leave the sleighs out in the bush where they worked. They'd sooner leave them in the bush because walking a mile in the morning would be far better than riding a cold sleigh.

I used to have five or six teams, and I used to hire for the sleigh haul, and even hire some for skidding. Farmers figured on making their living by going to the camp in the wintertime. I used to try to buy good big heavy horses. No use buying old horses, because they got old quick enough. I'd get a carload of oats and two or three carloads of hay in big bales, anywhere from 150 to 200 pounds. I used to feed a lot of grain. I never figured on feeding the horses too much hay, because the grain was far better for them when they were working. Lots of people used to cram the horses' mangers full of hay at night, but I used to tell the lads not to do that, just to feed them a flake of hay between the team, and that was plenty for them. Maybe they'd get stalled and wouldn't eat their oats at all. I've seen teamsters, even if the oat box was half full, they'd dump in another dump till they'd pretty near have the box full. They should have slacked off; they shouldn't have given any oats when they seen oats in the box. If you fed them too much, they wouldn't have an appetite for oats.

I never took hay to the bush. Lots of them did take a flake apiece, but I never allowed them to take any hay to the bush at noon, just oats and bran. Saturday nights I had two big kettles that they made [maple] syrup in, and I used to have the choreboy boil them full of oats for the horses. Boy, the horses liked boiled oats. We used to put some linseed oil in too; it kept the horses in good shape.

Most of the camps kept aconite around. It was good for horses with a bellyache. It was liquid. You'd give them half a spoonful, and if they were very bad you'd give them a spoonful. It would work very quick; in an hour they'd be getting better if they were going to get better. Then they had a medicine they called Bell's Medical Wonder. It was real good stuff for horses that were sick. Good for pretty near anything that was wrong with a horse. Just pour out a spoonful and give it to them. It was better stuff really than aconite, it worked quicker. Some of the teamsters used to give their horses arsenic. But a horse didn't need it if he was fed right, and it didn't help the horse anyway. It was just the imagination of the teamsters, throwing that stuff in them. I worked one winter for the Munn Lumber Company at Rock Lake, and there were a couple of guys there pretty near poisoned a horse with that darn arsenic. They never let on what was doing it. They made sure the bottle was well hid if anybody was around the barn. But if a horse was looked after and fed right, he didn't need none of that kind of dope.

GEORGE BRUNNE (b. circa 1910)

Did you ever hear tell of the Wild Wagon? A cadge team pulling a wagonload of pigs from Loring to Kidd's Landing ran away and dumped the teamster out on the road. That was the last that was seen of the horses, the wagon and the pigs. On certain nights for years after you could hear that wagon bumping over the rocks on the depot road. You still can. One night a fellow from Loring stopped to wind his horses and heard a wagon coming. He pulled off to the side to let it pass, and it kept coming and coming — but it never came. Some say it's the water falling over Dollar's Dam, something about the way a northeast wind hits the curtain of water. Anyway, you can hear it all the way to the [Ontario Lumber Company] depot campground two miles away.

ROY COCHRAN (b. 1905)

They used arsenic and antimony and all different kinds of things [to dope horses] and ruined them. I wouldn't do that. We had one tonic that was a wonderful thing, and I always meant before old Jack [Campbell] died to get the name of it. It was a powder from the drugstore. It wasn't a dope, it was a tonic. We never were told what it was. It helped the horses like you would take a tonic for a cold. Jack only allowed a few of us good teamsters to use it, but we were working our horses that much harder. I was classed as a good teamster because I worked my horses hard but I never overfed them, never allowed them to leave a bunch of hay in their mangers. And my horses fattened every day.

You could buy all kinds of stuff back in those years; there were no restrictions on it. You'd get arsenic in a little bottle. You gave them just what you'd hold on the tip of a knife blade. A good teamster didn't need to use that, but quite a few did. The horse did fine while he was getting it. It made him lively. He'd pick up and be in good shape, but as soon as he'd quit getting it his hair would stand on end like dry sticks. The horse would fail, go downhill, and it would take him half the summer to get it worked off. If a guy would go away and not tell you he was feeding this, you knew within a week, because the horse would get lazy and stupid. It took the life away from him. Some guys that drove team fed them Spanish fly. (Spanish fly was a thing that people years ago got into a lot of trouble with. It was a thing that worked on people, made them sexy. It came in a powder. Guys used to buy it, put it in chocolate and give it to girls.)

I understood horses inside and out. If a horse got sick on me, I knew what to do. In at Island Lake there was an aged mare, a very good one, and she got sick. We gave her aconite — we used to use aconite and laudanum together, but you can't buy laudanum anymore. It was used for colic, or something like that, to deaden the pain. Laudanum was a quieting thing, and so was aconite. They'd gave her the second dose of this and decided she wasn't getting better — because they weren't giving her time. They decided to give her this other dope we had. This would have been all right if they had started out with it in the first place, but they'd started with aconite and laudanum. I said, "Don't you do it; if you do you'll kill her in ten minutes." But they did, they changed, they gave her this dope, and it wasn't more than ten minutes till she gave one big rear and down she went. They wouldn't listen. When the two contacted it made a poison. It killed her.

They used to have a lot of trouble with blackwater. The neck of the bladder won't open when they get that. It keeps backing up to their kidneys and into their front parts and paralyzes their whole body. So you have to open it. Your finger and thumb, in a mare, will open it, but a male horse you have to use a small rubber tube. I've been there when they died right under the vet's hands. They didn't know, but I did. I found out I guess it's fifty years ago. I've seen horses that were down, paralyzed from the hips for as much as ten hours, and cured them.

Another thing was this Clyde itch. It would break out in their legs and they'd get rotten, stink. The horse couldn't stand still. It would just stamp and stamp. You'd think they were step-dancing. Especially if you put them in a new stable. Holy bald-headed, you couldn't sleep for them. A new stable, the sound was so much heavier; the floor hadn't got saturated. I've cured horses of it in a month. I bought the stuff out of the drugstore and it cost me bugger-all. I cured horses so that you wouldn't see a sign of it for two years after. I bought saltpetre, linseed oil and bran.

Percherons used to have a habit of getting cracked heels, big cracks across the backs of their bare legs you could lay a finger in, maybe four or five of them above the fetlock. They would bust open and bleed. I used the same stuff. They had tried everything — Gillett's lye, everything — but on the outside. Nobody learned that it had to come from the inside, that you had to drive it out from the inside. What you did, you cut the oats down a bit and gave them some bran in the place of it. You gave them the linseed oil as a laxative, and the saltpetre worked on the liver, the kidneys and the bladder. I could take Percheron horses that would have these cuts bust out on their heels and heal them up till they'd hair over and there wouldn't be mark. There was only one other man I know of that knew about this, and he's dead. You fed them bran every night, a cupful of linseed oil three times a week and a tablespoonful of saltpetre every second night for eight or ten days, then don't give them no more. Start over again, and the second round, you'd neither have Clyde itch or cracked heels.

As far as I'm concerned, a cross between a Percheron and a Clyde or a Percheron and a Belgian was the toughest team in the bush. A Clyde was washy, soft, it wouldn't keep up in the flesh. That fall we were in at Partridge Lake, there was this pair of browns. Oh, they were a torn-down pair of harum-scarums. The biggest one bust himself, he busted his insides drawing and died right on his feet. So after I had cadged the hay and oats and everything across the lake in the early fall, nothing would do but I'd take this [remaining] little brown and put it with one of the greys I drove. He was nerved up so bad that he was in white balls of froth. It took me four or five days to calm him down so he quit sweating. I never spoke ugly to him or was cross whatsoever. And when I hit the sleigh haul he was round as an apple. He couldn't be beat for that size of horse. When I came off the sleigh haul my team was still in good shape, where the other horses had failed. Mine failed a little, they expected that, but not much — on a four-mile haul where you decked your own loads, up to fifty logs in a load, and made four trips. We had ten teams in camp and three were pretty near tied. My team only weighed a little over 2,800. One team weigh 2,720 and the other around 2,730. But we had teams that weighed 3,300, and any of our teams could hitch onto their load when they were stuck and walk it out like nothing. If a load didn't come when they took a draw and you stopped them, why look out the next time because there was something going to go — whippletrees, doubletrees, something went the second time.

GOWAN GORDON (b. 1909)
You never used side straps or britchings or pole straps while skidding; you took them off because you didn't need them. Some people used eveners, a straight piece of timber like a doubletree with a trace chain on each end. In the middle they had a roller and one long trace chain hitched to the two inside traces of the horses. No matter what angle they went, both the trace chains were tight and pulling. And an evener was a lot lighter than a doubletree and set of whiffletrees to throw around in the bush. The only thing was, if you got in a narrow place it was longer than a doubletree, about five feet long it seems to me. You used tongs on the bigger logs, and a skidding chain for the smaller ones where you could bring in two at once. One thing that was bad was catching on a limb and pulling it ahead, then it would come back and take you across the legs. I had my feet lifted right out from under me one day. And you had to watch all the time to keep on the right side of your log, on the outside of your turn. If you were going around a right-hand corner you had to jump over and be on the left side, where it was going away from your legs.

ALBERT SCOTT (b. 1895)
Standing at the anvil doing all the pounding, that's the big job [for an apprentice black-smith]. That's why they liked to have an apprentice working two or three years for nothing — and to cut all the clinches, pull off shoes, dig the dirt out of [hooves] and do the paring down and fitting of the shoes. After you're there about a year they might let you learn how to drive a nail. When I was learning you welded a caulk on the toe. You turned the two heels up and made [heel caulks] out of the shoe itself, but the toe caulk, there was a space where you welded a caulk. If it was spring or summer you'd get what they called a mud caulk. It was about half an inch wide and three-quarters of an inch high, and it had a little tit on it. Those were handmade. You just welded them on with a hammer. After a while they started making them, and they made both sharp and dull caulks. Sharp ones for winter. If [horses] are working in the bush they have to be sharp shod in order to get a hold on the ice roads. The old blacksmiths used to weld on dull caulks, then draw them up sharp. They sold you a long block of iron with slots cut in it to fit the different sizes of caulks. Then they started getting in patent caulks. A factory-made sharp toe caulk was about an inch high, half an inch wide at the bottom and tapered to the edge. It was made out of iron that was cut sharp, not pounded sharp. They made them in different sizes. A five caulk was nearly four inches long, but that was for a horse that wore an eight shoe. That's a big shoe; the average big shoes was a seven.

Clydesdales and Percherons were a lot easier to shoe than the average mongrel horse. They're more lifey and they're harder to get along with. You take any colt, if you start picking up his feet when he's young, when he's a colt sucking his mother, you'll never have any trouble shoeing him. But if you let him grow to be two or three and he's never had his feet off the ground, I tell you, you got a contract on your hands trying to shoe him. And these damn western horses, when they started rounding them up on the prairie and chasing them into corrals, loading them in boxcars and bringing them in, they never had a halter on in their life, and some of them were six, eight or ten years old. Nobody knew how old they were.

Operating a crazy wheel.
— CHRIS WATTS

Key Valley Railway locomotive hauling logs to the Schroeder Mills sawmill at Lost Channel, about 1925. — JIM LUDGATE

CHRIS WATTS (b. 1911)

In my experience you can log through a bush with horses and you're not doing any damage to the bush. But go in with big machines, the way they do now, and you spoil more timber than you take out. On the other hand, I was glad to see horses go out. I was always fortunate I worked in camps where we had good horses, good feed, good harness and good teamsters. If they had a teamster who wasn't good, he wasn't there long. But there were other camps in the country where you heard of them beating horses to death, not feeding them properly and asking them to do twice as much as they were able to do.

Dan McKelvey, Mark Taylor's bush foreman, was a hard man on horses. Taylor told us, "I put horses here to work, but a crippled horse is no good to me." One time we were between High Lake and Balsam, and we had a vee-plow breaking the first frost. We came to a creek with about three feet of water in it and an inch and a half of glare ice on top. There were a couple of guys with me, and I said, "Break that ice with the axe." Dan says, "Put the horses in. They'll break her down." I said, "No, they're only going to cut their legs, And if they don't go in they'll slip on it. They've only got summer shoes on." He says, "Aw, that's what they're for." I says, "Here's the lines; if you want them in there, you put them in. I know Taylor don't want to hurt the horses." Dan was a nice old fellow, but he didn't have too much sympathy for a horse.

You get attached, you love those horses. We'd go to the barn at night and curry our horses, and all the teamsters would be telling stories. Everyone skidded the most logs in one day and the biggest. First thing you did when you came in at night, you took the harness off them, put the blankets on and gave them a bit of hay. You went in and washed up, had your supper, then went back down and gave them their oats. You took the blanket off one horse and you'd curry that horse, brush him and curry him, then blanket him up again and do the other one. And you petted them. When you walked in the stable they'd give a little whinner, glad to see you come in. And they always seemed to like to go to work in the morning. They *wanted* to go to work in the morning. Horses, if you feed them right, harness them properly and use them right, they like to work. I never was unlucky enough to work in a camp where they were rough on their horses. They always supplied the best — molasses to give them in their oats at night, boiled oats every Saturday night with linseed oil. The choreboy would boil two large cast-iron kettles of oats, and after they cooled he would put one gallon of raw linseed oil in each kettle and stir it real good. If you put the linseed oil in the oats when they're boiling, you boil the oil and then it constipates them. I remember we had a new choreboy and McRae forgot to instruct him how to do it. He boiled the linseed oil with the oats. McRae just happened to come down before we came in and asked him whether he put the linseed oil in, and he said, "Oh, I put it in before I boiled them." Well, he had to dump the whole works out and we had no boiled oats that weekend because it was too late. They had to start in the morning to boil them.

February 8, 1878: No sleighing on the main roads and very little in the woods. The sawlog crop will stand a poor show of being watered this winter. — D.F. Macdonald

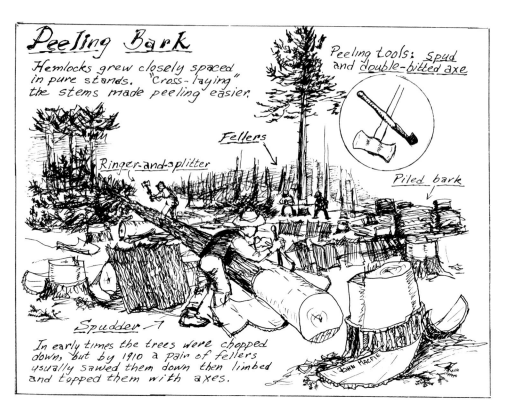

Peeling Bark

Hemlocks grew closely spaced in pure stands. "Cross-laying" the stems made peeling easier.

Peeling tools: spud and double-bitted axe

Ringer-and-splitter

Fellers

Piled bark

Spudder

In early times the trees were chopped down, but by 1910 a pair of fellers usually sawed them down then limbed and topped them with axes.

JOHN MACFIE

Hemlock operation. The trees were felled and stripped of their bark in June, and now in Autumn, the trunks have been cut into sawlogs and piled in skidways. They and the bark piled on the right will be hauled in winter for the Lake Rosseau Lumber Company.
— BARBARA PATERSON

68

PEELING AND TANNING

GUY SMITH (b. 1885)
They used to peel all the hemlock; they used it for tanning. The Moon River and all through that country was pretty near all hemlock, and that's where the Conger Lumber Company got most of their hemlock. The bark was sent to Bracebridge, where they had a big tannery by the river. It takes a lot of water for that job.

Peeling starts in the spring. As soon as the frost comes out of the timber the bark loosens, and it lasts until maybe the middle of July. In the fall they'd cut them [the peeled trunks] into logs and skid them, and they'd drive them next spring. Lying in the bush all summer with no bark on, they were nice and dry and floated real good. Too good. When they had all peeled hemlock in the drive it was pretty smart stuff, I'll tell you! It was light in the water and with no bark on it was pretty smooth. It floated quite high and was very, very quick. You had to watch yourself on it. But I liked driving it, because if you wanted to go across the river you could get onto a log of peeled hemlock, give yourself a shot and you'd go halfway across. Give it three or four paddles with your pike pole and you were on the other side.

ART DOBSON (b. circa 1905)
Bill Taylor had the tannery in Parry Sound. It was a frame building with a lot of cement work inside — big vats. Some [were] ten feet long and six feet wide, and six or eight feet deep. There was a ground floor and an upper floor, and a big furnace. They had a big supply shed for tanbark. It would hold twenty or thirty cords. When they wanted to use it in the tannery they put it in a hopper and through a grinder to ground it up fine. There was terrible dust from it. When it came out it was ground up in pieces the size of your finger. The tanbark went into the vats, in the liquor. They had a high pile of beef hides with the hair on, each one with a layer of salt on it. They put them in this vat to take the hair off, put them in on long poles with hooks attached. They'd leave them there for quite a few hours, then put them in another vat with something different to cure them. There was a great big drum about three feet across and eight feet high, and a door in the centre of it. They used to put some kind of grease in, throw the hides in and run them around. It looked like axle grease. God, it was a mess. The hides were real soft then, the hair all off. They were a grey colour. When they took them out of the wheel they put them back on a table and there was a big cement thing that came down "whack, whack" and that shot all the grease out of her. Oh, there was grease flying. That bumped it right down till it was smooth. And when they took it out of that they hung it up in the upper floor to dry.

All these old lads buying cattle, and butcher shops, that's where they got their hides. Oh, it was a smelly place, stink like the devil. My brother Ed worked there till old Taylor died. He used to put the leather through all the treatment. He could cut laces better than the machine. He just took a knife and ripped her down to beat the devil, any width he wanted. I guess Taylor worked in the tannery when he was younger. The old lad was so used to walking around where it was slippery he didn't lift his feet off the ground when he walked, just trailed his feet, skidded them.

HECTOR WYE (b. 1905)
One winter I got a few days off school and I helped Ab Whitmell draw bark to [Boakview] station. There used to be a road down Shawanaga Lake for drawing bark and cadging. Pretty well all the farmers along here were drawing that year too. The Lake Rosseau Company bought the bark. They bought it in the pile where it was swamped out.

The bark was worth almost as much as the logs. Even the bigger companies always started a gang of men in the bush when the bark was peeling. When they felled a tree, if the axeman could cut the notch a little deeper at the centre, when they sawed it down, the outside corners hung on. If it hung up on the stump, then it was easier to peel. But it must have been annoying when the guy came to skid it. When they peeled the bark they just cross-

Piling tanbark.

Skidding hemlock logs at a Peter Company camp in October, 1910. The bark was removed from the logs in early summer for shipping to a tannery. Rolling logs up spiked skids had by then been largely superseded by the decking line as a way to build skidways. — PAC PA 16809

piled it in the bush in little piles around close. Then there was always the job in the fall of swamping the bark — swamp it out and pile it in piles along where the road would be, where you could get it in the wintertime. They'd use a sloop for this. It was still piled with the bark up, same as in the bush, but they turned it over to put it on the sleigh because it was hooped so that when they put it that way the outside [sloped] up. On a sleigh a cord of bark took up an awful lot more room. It all fluffed up. It wasn't a matter as a rule how much the horses could draw, it was how much you could get to stay on. Then, when it was piled in the [railway] car, it was piled bark-side up again.

MARSHALL DOBSON (b. 1892)
I peeled bark for Roy Macfie and I peeled bark for Johnny Simpson. Two men fell and limbed, one rung and split, and one man spudded. Four men in a gang. They'd average four cord in a day. Each man was supposed to be able to peel a cord. You'd start in June and go to July, maybe the tenth, and that was it. It was starting to dry up then. It doesn't peel good after that, you don't get a good turn. You'd start about four o'clock and pile what you peeled the day before. You'd cross-pile it, maybe four trees to a pile. For fly oil we used tar and hog's lard. If the flies didn't bite hard enough you burnt yourself with the tar.

Hemlock bark stacked beside the railway at Boakview. — BARBARA PATERSON

A sleigh load of tanbark.
— ART DOBSON

Camp cook (r.) and cookees. — GEORGE KNIGHT

A turn-of-the-century lumber camp cookery. — JAMES T. EMERY

CAMP LIFE:
COOKS, COOKEES AND COOKERIES

DICK BREAR (b. circa 1900)

When I first started to work there were no women cooks. Sometimes the food was pretty bummy, and sometimes it was good. The best fellow I ever ate after was Big Bill Macdonald from Orillia. That was up at Pakesley. Orv Gray was a boss up there and when Tudhope and Ludgate started to work down here [Maple Island] he was boss. He went into the cookery one day and said, "Bill, can you cook for a few more men?" Bill said, "Gray, I can cook for more goddam men than you can yell 'All Aboard' for!" We used blackstrap as much as they use table syrup today. Many the time you'd get that old side pork, and it was pretty well all fat. As long as it was cold I could eat it like cheese. Take a chunk of that fat pork and twist it around, put syrup on it, and down it.

NORMAN CAMERON (b. 1894)

The Turner Lumber Company logged down in Ferrie Township. They drove logs from Snowshoe Marsh down the Kimikong into Chartier Lake and down the Pickerel. Dad worked for Turners, and we were living at the old Gore place. Two cousins of mine, the Culligans, came down from Golden Valley to visit us in the wintertime, and we had a team of dogs, so we went in to Turner's camp on Snowshoe Marsh. That was the only thing we had to show off. The cook made nice apple pie out of dried apples. He'd bake apple pies by the dozen. He set pies out on the table for us, and I ate apple pie until I looked like an apple pie, and so did my brother Mel, and Dave and Charlie Culligan. We must have ate a dozen pies on that man. I remember at the last Dave had a chunk of pie yet, finishing it up, and there was one other piece left. The cook says to him, "You might as well take the rest," and pulls out Dave's pants and dumps the last piece down the front of his pants. The two Culligan boys were both shot in the First World War.

Jack Irish was foreman for the Tonawanda. He had a bad stomach; I remember my dad telling about it. He was sick of salt pork and he fished all winter through the ice of Irish's Lake, but all he was catching was little perch and chub. One Sunday he caught a little pickerel. He cleaned it up and took it in to the cook — they had a Frenchman for a cook — and he said, "Now you cook that for breakfast tomorrow morning." Monday morning Jack Irish came in and said, "Where's my fish?" "Right there, Mr. Irish, in the pot there." He lifted the lid and there was this white gooey thing. The cook had put the fish in a pot and boiled it, then taken some flour, mixed it all up and put this white gooey stuff on it. Irish says to the cook, "What do you call that!?" "Saucé la goo," he says. "Saucé la goo." Irish grabbed the pot and made for the cook. The cook made out the door and Irish threw that pot at him. It hit the top of the door and fell on the doorstep upside down. He says, "I'll 'la goo' you, you son-of-a-bitch!" He thought he was going to get a nice fried pickerel for breakfast, and here he'd made a French meal of it.

BERNARD MOULTON (b. 1894)

There was no fruit in the wintertime, no apples or oranges or bananas, nothing like that. Potatoes and meat. There was nothing very fancy them years. But you seemed to get along all right. You got paid when you quit, or the job was done. Doctor's fees were twenty-five cents a month. Macdonald's was the main tobacco then, Macdonald chewing tobacco and Macdonald pipe tobacco. As far as smoking cigarettes, it was Old Chum.

MEL CAMERON (b. 1892)

When my dad came here to work they used the camboose, an open fireplace, in the men's camp. South of the Pickerel River there was a camboose. The camp used to get pretty cold some of them severe nights. You didn't want to let your hair grow too long, for it would be fastened to the wall when you woke up in the morning. But them old camboose cooks, boy, my father and grandfather sure recommended them. They lived a lot plainer than we do, rough food but cooked good — beans and dandy bread, and they used to buy beef from the settlers if they could. And salt pork, long and clear. There was none of this breakfast bacon. But in our time they sent butter and pies and cakes out in the lunch boxes, and a lot of camps sent out hot dinners for the noon lunch. They ran what they called the "pung." The foreman used to clear out a place that was general around where his gangs were skidding, and they'd all come to this place. They'd bring this old pung up, a jumper with a box on it, and there'd be pots and everything in there — hot potatoes and beans and meat. You ate about as good a meal there as if you'd been eating in the cookery. They built two fires and they had logs all around it for seats. The dishes and everything were set out and you'd help yourself. Some of them used to help themselves too much. There'd be the odd greedy one who would clean up all the pie and sweet stuff. Never used to bother me too much. If they had good meat and potatoes, that satisfied me.

CHRIS WATTS (b. 1911)

The camp was about two and a half miles east of Kimball Lake. The teamsters had their own sleep camp because we were called earlier in the morning to go and feed our horses. The teamsters' camp was about twenty-four by thirty feet with sixteen double bunks made from round poles, one above the other. You put cedar brush on the poles then hay on top of the brush, then had a blanket over it. A bag of hay was your pillow. The large sleep camp was about thirty feet by sixty, about enough room for 110 men. There was one wood stove made from a large drum, with a rack made from small poles around it to dry clothes and rubbers. You can imagine the odour from the rubbers and socks. In between the teamsters' camp and the men's camp there was a washroom about twenty-four by twenty-four with a sink made of wood lined with galvanized tin full length along one side. There was a stove with a forty-five-gallon barrel backed up to it for hot water, and a barrel for cold water. Also there were washtubs and stands and washboards you could use on Sunday to wash your clothes, but not many bothered. There were about four grindstones in there too. Once every couple of weeks they'd want to sharpen their axes, and if they happened to hit a stone they'd sharpen it at night. They carried a little whetstone in their pockets. Those axes were just like razors.

The cookery was about twenty-four by sixty. Four dining tables sat about thirty men at each. Two large wood-burning cookstoves, two bedrooms and a large kitchen. There was one head cook, one second cook, two cookees and one choreboy. Percy Roe was second cook. He went out at Christmas to get a bunch of liquor — after that he got converted and never drank, but at that time he did. He went out for Christmas and brought in a bunch of booze. And he brought a pair of ladies' shoes. He only wore a size four and a half, and at night when everybody was in bed he'd go out, walk around the sleep camp, over to the office and around, then put them away. About every two weeks he'd go out at night and make these trips around. In the morning the guys would all be out there whinnerin' and whistlin' and lookin'. And George McRae would say, "There's a woman in here someplace, and no woman has no business in this place." He was looking all over the place trying to find this woman. I think there was only the cook and Percy and I knew about it. There was a lot of fun had.

The most important thing about the men's camp is an outside toilet built of logs about twelve feet long and six wide, a long pole about six inches diameter to sit on and another pole for a back rest. Toilet paper was whatever you could get: newsprint, boxes, books. Some of us took toilet paper with us.

Bedtime rules were that the choreboy would come along and say, "Ten to nine," and at nine p.m. all lights were out. If anyone talked to his bed chum after nine you would find a boot thrown at you. Except Saturday night, when you could stay up until ten o'clock. The only time you saw the camp in daylight was on Sunday. All we had for light at night was the coal-oil bracket lamp. All teamsters were supplied coal-oil lanterns. The choreboy kept

In a lumber camp cookery. Note the loaves of bread set to rise in the warmth above the stove. — GEORGE KNIGHT

lanterns and lamps clean and full of oil. When we went to work in the morning we left our lanterns burning. The choreboy would put the lanterns out after we went to work, and he cleaned the stable.

I could cut hair. Every Sunday I would cut about eight or ten heads of hair. One package of tobacco for a haircut. At the end of March, when leaving camp, I had enough tobacco to last all summer.

WALTER SCOTT (b. 1893)

There used to be some great camp cooks. Others weren't so good. There was an awful difference in the way cooks put vegetables and things on the table. One fellow who was a real good cook was Billy Roach. And another was Johnny Jackson. Some cooks are very expensive because they're wasteful, but nothing ever went to waste in his cookery. Any bread that got stale went into bread pudding or something like that. He could board men cheaper than pretty near anybody.

We used to get in a good supply in the fall of pretty near everything we would need through the winter — sugar, flour, lard and butter, anything that would keep well. We'd build a roothouse to keep the stuff in. When you get away back it's pretty hard top get potatoes and vegetables in the wintertime because they freeze. Lots of places they kept pigs. Manley Chew had different camps around, and he used to raise enough pigs on Franklin Island to do him all winter. Maybe he'd have 100 or 150 pigs, and they'd make their living in the bush, just run loose all summer, then he'd kill them for his camps in the winter. He had camps up that way, and he used to keep these pigs on the island because they couldn't go anywhere. There are lots of fern brakes on the island, and there's nothing pigs like better than them old fern roots.

MAY VOWELLS (b. 1889)

I was born in 1889, and I'd have been twenty-two or something like that when I first worked in a lumber camp, at Pointe au Baril for Graves and Bigwood. My father-in-law, George Vowells, had the job. Our camp was on the shore of Georgian Bay, halfway between the village and Oldfield's tourist house. The first while there were about ten men, and the last winter we had about twenty, I believe. I had my sister-in-law with me to help. She was only around sixteen or seventeen, but she could peel potatoes and help around washing dishes and things.

We'd get up early, five o'clock maybe. I'd make porridge and that, then the men would take a lunch to work. We'd put up a lunch for them, slice meat and put it and bread in a basket. The men would come in for supper and we'd have a regular dinner then — potatoes, meat, vegetables, cookies and pie and everything.

It was just a rough camp, and cold. The men had their part, and there was a little room between it and the cookery. One time two men came in and Grandpa hired them. We didn't know them or anything, but they started acting up with my sister-in-law, talking kind of nasty, dirty-like to her. At last they went out to the men's camp, but the next day they didn't go to work, and we were there alone with them two men. I locked the cookery door, and when it came dinner time they wanted to come in and get their dinner. I said, "No, you're not coming in here!" I wouldn't let them in. So they got mad and went away.

CAMP LIFE
SINGING AND FIGHTING

JIM McINTOSH (b. 1896)

I'm eighty-five years old and I never knew what a holiday was in my life. I never had one, I never wanted one. I never drew five cents of unemployment insurance nor five cents of compensation. I never went on relief, I never asked for nothing. I was never out of work in my life. If I couldn't find a job for myself I'd grab a saw and go out to the bush and cut ten or fifteen cord of wood and sell it. I worked for $12 a month, sixteen hours a day, and raised two boys, clothed them and educated them.

I often wish I was thirty years old again. If I was going to live life over, I'd want to do just what I did before. I loved the bush. I have gone in on the twenty-second of July, come out on the last day of March and never was away from that camp. Washing clothes was the only entertainment I seen — in a couple of great big kettles out in the yard. An odd time somebody might bring a violin in after Christmas. Saturday night somebody'd sing a few songs. Charlie Fenton from Sebright had the loveliest, softest voice. His voice would carry; it was low, and it just seemed to roll out so smooth. There was old Bill Wilson, Jimmy Campbell, and another fellow from Sebright, old Charlie Baker, boy oh boy, he was a good singer. In 120 men you'd be surprised how many could sing. On Saturday night everybody would sing all the songs they knew. I knew fifty-two songs. War songs, true love songs, shanty songs, river-driver songs. I'd hear a man sing a song a couple of times and I'd have her.

[He sings]:

> Oh come all you boys if you'd like to hear
>   how I got up to the bush last year.
> Unto a place you all do know,
>   it's a great big lake called O-pe-on-go.
>
>   *[Refrain]*
> *Come a-ram-tam-tam-ta a folly-deedle-da, ram and a roar and rum-come-away.*
>
> Oh from Arnprior we did set out
>   with old Jack Brant for to show us the route.
> We marched from there unto Renfrew
>   and there we met our gallant crew.
>
> So to make acquaintance we all did begin
>   but some did dip too deep in the tin.
> And some jolly lads got on a spree
>   and to hire a rig they did agree.
>
> Oh I'll say by God didn't we feel big
>   into a nickel mounted rig
> And for Dacre town we hoisted a sail
>   and we all sat there to the Prince of Wales.
>
> Oh Spratt came out for to welcome us in
>   and he laid down both whisky and gin.
> The landlord's treat was a merry round
>   and we drank our health to Dacre town.

So it's now we are up to the O-pe-on-go
  to toil all winter through frost and snow.
Jolly lads though we may be
  it is in the springtime we'll get free.

Oh the first lad came with the foreman's team,
  you would swear by God they were drove by steam.
Five-trip haul on a four-mile road
  and forty-five big logs to the load.

Oh the next lad came with a span of blacks
  and they were the boys you couldn't hold back.
Out in the morning feeling fine,
  and in at night away behind.

So it's now we're through with the O-pe-on-go,
  we toiled all winter through frost and snow.
And Jolly lads we still will be,
  for in the springtime now we're free.

A fellow said to me the other day, "You were a bugger to sing when you were young." I said, "Well, I was a bugger whether I sung or not."
[He sings]:

I'm a cattle-buyer from the Burks Falls town,
  and I drove from the break of day.
As my buggy whirls by you can hear people say,
  "What's the price of cattle today?"

  [Refrain]
  So we'll all go over into Ryerson,
    and down into the Township of Spence,
  Up to "The Mag" and home with a jag,
    and to hell with the enormous expense.

Sure he drove a little roan, he was all skin and bone,
  and the buggy with the three wheels on.
Now I'd far sooner walk than listen to your talk,
  so I said, "Old nag, punch along."

Then I met a young lady on the Miller swing bridge,
  she was waiting for the boat *Glen-o-ro*.
She was going to be wed to Charlie, so they said;
  he had gone for a new suit of clothes.

And I met old Billy Grant on the corner of the road,
  and he said, "How many did you rob?"
Now I haven't bought a thing since I left Kit-er-ine,
  but I'm still going out on the job.

Then I met old Mister Brandt way up in Gollyer's swamp,
  and it seemed to be upon election day.
By the golly darn sure he wouldn't vote reform,
  for he didn't like the reci-procity.

*Interior of a circa 1925
sleep camp on the
Schroeder Mills limit.*
— JIM LUDGATE

*A circa 1905 camp near Dunchurch. Foreman Charlie Piatte is wearing a fur coat, and the
hatless man seated just to his left is James "John Dollar" Madigan, renowned for his irreverent
humour.*

Bill Wilson made up "The Cattle Buyer" in 1912. He came back to the camp and told about the sinking of the *Titanic*, and he sung that song for the first time. Do you know the Wilsons who used to make the homebrew out of Magnetawan? Josh and Jack and Bill. Three awful men. Great lumberjacks. Do you know where the Miller swing bridge is, half-way between Burks Falls and Magnetawan? That's where they were born. The passenger boats used to go through there. The old *Glen-o-ro* used to run from Burks Falls to Magnetawan, and that was a swing bridge for the boat. It don't swing no more.

The lumberjacks used to get me to take ten and twelve bottles of their whisky back into the bush for them when I run camp. They'd give me the money, and I'd go down and buy it. For $2 a bottle. Made out of rye grain. Oh, it was strong. I saw them, the first that came, put a drop there, a drop there and a drop there, then light this one and the whole three would light. It was that strong. Oh, you couldn't drink the first that came at all. You had to run it until the liquor got weak, and mix the whole thing. Then they'd take brown sugar, burn it real black, and put a little whisky in that and shake it. It would colour that just perfect. Why, you could make rum, you could make anything at all out of the same brew with the flavour they put in. They had her rigged up, all right. It was in a hill. They dug a hole in a sand hill and built it up with cedar timbers. You could walk right by it and never see it. They had a bunch of four-inch pipes that went down underground into a creek, and the smoke went into the creek, and the speed of the water going took the smoke. You couldn't see a pick of smoke at all. It was the First World War, the last time I was down there. They were three wonderful men, wonderful singers.

Another song I could sing was "Camp Number Three." I worked in Camp Number Three for years. It's exactly ten miles from here [South River], the other side of the Big Spring.

[He sings]:

> Come all you lumberjacks and I'll sing you a song,
>   a nice little ditty, it won't take me long.
> It's all about lumberting, you plainly will see,
>   one winter we spent down in Camp Number Three.
> We had fine times in Camp Number Three.
>
> We had a nice foreman, a cute little man,
>   he tried to get out all the logs that he can.
> But he hadn't no money, I'll tell you the cause,
>   a little too often he diddled them squaws.
> A fine man for Camp Number Three.
>
> Oh we had an old blacksmith, you all know him well,
>   the god-damnedest blacksmith on this side of hell.
> He burnt all the steel, it was like charcoal,
>   and his hooks wouldn't catch in a greyhound's arsehole.
> What a fine man for Camp Number Three.
>
> Oh we had a fine team called Paddy and Queen,
>   the god-damnedest team that ever I seen.
> They'd deck any log, 'twouldn't matter how high,
>   the hardiest was Paddy; he'd deck her or die.
> A fine team for Camp Number Three.
>
> Oh we had an old cook, the daisy wee thing,
>   at the wee hour of daylight he'd give us a ring.
> At the wee hour of daylight he'd call us to chuck,
>   "Get up, you sons-of-bitches, it's broad daylight!"
> What a fine man for Camp Number Three.

Oh we had a fine clerk, that daisy wee thing,
   he'd scale all the logs on the dump in the spring.
He'd take out his rule and go scale every log,
   but he didn't know punk from the arse of a dog.
What a fine man for Camp Number Three.

Oh, I sit here and think of the good old days. I wish they were back again. I wish I had the years that I logged for myself. I wish I had the same men as I had. It was a pleasure to get up in the morning and go to work with them. You didn't have to tell them nothing. They knew what to do, and it was done.

January 17, 1917: They buried young Carson, who was killed by the limb of a tree in Burpee, in Bracebridge today. — D.F. Macdonald

JACK CHISHOLM (b. 1896)

There was one thing about a lumber camp, I never saw one yet that didn't have a violin in it someplace. There was a big fellow by the name of Jim McDonald when I worked up at Ardbeg, he was good on it. And there were a lot of them good on the mouth organ too. We used to have stag dances on Saturday nights. Oh yes, you always had a little enjoyment on the weekends. The other nights we felt like going to bed.

You always had friends, and I tell you, it wouldn't pay for the other guy to pick on a boy, because they'd get all the pickin' they wanted before they were finished. You always had friends in a camp. The main thing with a boy was not to have too much to say. If you got too mouthy nobody liked you. If a question was asked, answer in a civil way and you'd always pick up a few friends. I know there was one fellow took my part one night. This guy was going to slap my ears. He said something and it referred to my mother. And me, quick tempered, I hit him. 'Course I wasn't man enough for him, because he was a solid chunk of a man, but my friend took it up. The guy didn't bother me no more.

MEL CAMERON (b. 1892)

Bob Montgomery was a tall man. You didn't tackle him unless you were pretty good. Not very many beat him. They say he took an awful beating in Parry Sound one time though. He had a brother who ran a hotel on the harbour side, and he got into a fight and gave this Italian quite a licking before they got him stopped. This Italian went and trained, and he came back and met old Bob again, oh, a couple or three years afterwards. And boy, he set him back on his haunches pretty good. He'd trained, you see. He fixed Bob.

One time I was at the Wildcat they [Montgomery and others] got into an awful fight, four or five of them. They got out of the hotel and out on the grass, and it's a wonder some of 'em didn't go into the river — it wasn't very far to the edge of the river there. But anyhow they got it settled. And the roars of him! They called him "Roarin' Bob." You could hear him! You'd think somebody would be killed, the racket till they got him quieted. Around Dunchurch there's guys will tell you if he went down to the hotel there, he got into a fight every time he hit Dunchurch. He fought with everybody he could fight with, yet he seemed to be very friendly in a lot of ways. I knew the Montgomery family. Mrs. Bob Montgomery seemed an awful nice woman.

BOB GIBSON (b. circa 1892)

There was a liquor hotel in Parry Harbour, a liquor hotel in Waubamik, the hotel in McKellar, the hotel in Dunchurch and one at The Wildcat. At The Wildcat they got the logging days business. My uncle, "Roarin' Bob" Montgomery, ran the first liquor hotel in Dunchurch, and they had the one at The Wildcat too. Joe O'Connor, he was always out for fighting, and so was Roarin' Bob. They got into a fight right in the sitting room this night. They were pretty near tied. O'Connor got a hold on Bob, and Bob couldn't break his hold. The big box stove was there, with a fire in it, so he kept working away, and O'Connor never

McDonald's hotel in Ahmic Harbour with the Graves, Bigwood Company's storehouses on the waterfront. — ALBERT STUCKEY

Art McDonald, proprietor of the hotel in Ahmic Harbour.

The Montgomery House Hotel in Parry Harbour. — PARRY SOUND PUBLIC LIBRARY

noticed what he was doing. Finally he got him up against the box stove, and when he started to sizzle, why he let him go. Another time they got into a fight in the yard, and Bob's little boy Randolph was there. Bob had O'Connor down and Randolph said, "Hang to him, Dad, hang to him. I'll run and get the axe and cut his head off!"

They used to walk from Dunchurch to Loring in order to give a lad a trimming. He would be talking fight to somebody and telling how he could do this other fellow, so the fellow from here would walk clean to Loring to give him a licking. They didn't do it all with their fists. I've known them when they were drawing bark to take the stakes out of the rack, ironwood stakes four feet long, and fight with them till they knocked somebody over.

ARNOLD McDONALD (b. 1907)

There were a hundred men in Ludgate's camp. The two sleep camps were built of logs, and a wash place in between. The Polacks slept on one side and — I don't know what you'd call us — we slept on the other side. But we mingled back and forth. Them foreigners could sing damn good. They had an accordian. We didn't know what they were singing of course, but it was nice to listen to them. Charlie Baker was the foreman and he was a real good foreman. He knew a day's work. Us fellows that drove team were up an hour before anybody else and we had an hour more work at night. Through the week — and it was proper — every night at ten to nine the bull cook, the choreboy, came around and yelled, "Ten to nine," and you went to bed. He put the lights out at nine. The sun was wasn't up when you got up, and you were working pretty hard, but us young lads were always full of ginger of course. Some of us would be talking a little bit and Charlie would be listening outside. The next night he'd come in there and pretty near knock the door off and bawl us out. "You young buggers, if you don't want to go to sleep, keep quiet. There's older men here that wants to go to sleep!" And he generally mentioned my name. During the week he wouldn't allow any hollering. He wouldn't let you play poker for money. You'd play for matches. They could play music; there's lots of talent in a hundred men. Baker came from Sebright and there was different ones from down there, "down-homers" we called them. Different ones of them had guitars and violins, and back in them times there were some pretty good old-time singers.

On Saturday night we'd have a stag dance until twelve o'clock. Baker brought in pails of apples and stuff from the cookery. You could squeeze a couple of songs out of him. And there was Jack Montgomery from Sebright, he was a good singer. And the gang at Tudhope and Ludgate's mill, there were a lot of pretty good singers there. Chris Carlton, oh god, he could sing a song that would have tears running down your cheeks, then he'd turn around and sing a song so dirty you could smell it! And Con Ross from Magnetawan, he was a great old-time singer, he had some good songs. And my brother could sing a good song. Jack Delorme, he worked in the lathe mill, was he ever a good step-dancer. He had so many changes and steps. Of course you've seen Ab Whitmell step-dance, and Ernie Carlton. Ab Whitmell used to step-dance the "Twelfth of July." He always took the prize. He danced heel-and-toe quite a lot. Wes Aulbrook was a step-dancer, and old Frank LaFlamme that run the dance hall at Ardbeg, he could step-dance pretty good. He was a piler at Tudhope and Ludgate's. It was nice to walk through that yard and see his lumber piles, they were put up so nice. He'd never stoop with a board. He'd just put the other end down, and when he let her go his foot was on her and she stayed there.

DON MACFIE (b. 1921)

Some company got timber rights around Debow's Lake, but there had been some rivalry about who would get them. This company got Jim Canning and a helper to do some cruising and boundary-fixing in the fall. They kept a ten-pound pail of corn syrup at their dinner place. After they had eaten it down some, the helper noticed something in the bottom of the pail and jokingly remarked that it looked like a turd. After a few more days, when they were down nearer the bottom, Jim took the syrup paddle and dug it up. Looking closely he said, "Well, by Jove, it *is* a turd!"

Jack Moore told me this. One time he, Bill and Marshall Dobson, Dick Cooper and some others were in the lumber camp behind Eagle Lake. At night there was usually a little hellery, with everybody trying to get one on Dick, who was strong enough to put them all in a heap at once. One night they got a hoop off a butter barrel and started to see who could step into it and work their way through it the quickest. Dick came along and fell for the bait. "By the gee whiz, I can do that!" So they let him have his try. When he got the hoop up where his arms were pretty well pinned, they all jumped him and rolled him around and around and end over end on the floor, with Dick all the time telling them what terrible things were going to happen to them if he ever got out of there. Finally they had their sport, and they all lit out the camp door, everyone in a different direction. It was a moonlit night and maybe twenty below. Jack took off down the log road to Whitestone Lake in his sock feet, with Dick after him in his bare feet. Dick caught him and started back to the camp, systematically reducing him to a heap of misery. When they came to the first bunch of horse manure he made Jack eat some, and after going further he opened his shirt and underwear and started filling him up with snow, shaking it down like a bag of potatoes until he was filled to the neck. Then he let him go and went back to camp. He waited until the rest had to come back or freeze, and dealt with them.

DICK BREAR (b. circa 1900)
My brothers, Walter and Johnny, used to take liquor, booze, away up around the Kimikong. They had four dogs, a double team of dogs. There were camps all through there and away northwest of Simms Lake. They ran there for quite a few years, mostly on weekends. You could buy [liquor] then at Dunchurch. Old Newby run the hotel there just before you cross the bridge in Dunchurch; he had an iron paw.

When the sleigh haul was done they came out to the Wildcat and stayed till the river drive started. Spent their money. They'd go on the drive, and when the drive was done they'd come back and stay at the Wildcat or McAmmond's place till the bush work started again. There used to be Mike Folliet, Dave Gordon, Freddie Connor, Eddie Connor and Big Dave Connor — three Connors. That was about the bunch. They used to be fighting all the time. Parrish ran the hotel at the Wildcat, and when he left there he ran the hotel in Loring. Montgomery had the care of [The Wildcat] one time too. I heard my old man talking about one time he was up at Magnetawan and there was a bunch of Gypsies. Somehow they got into an argument and the old Gypsy took Montgomery by the neck and the ass of his pants and put his head in between the spokes of the wagon [wheel]. He shoved his neck down next to the hub, where it was narrow.

ROY SMITH (b. 1912)
I got this story from an uncle of mine, Dave Gillespie, who came to Parry Sound to work in a lumber camp as a young man in the late 1800s. He arrived [by train] at Utterson and walked to Christie or McDougall Township. A number of boys had been hired by some foreman or organization away from here to start this new camp. When he arrived, there were some men outside the bunkhouse running around in a circle tramping down snow. Being a newcomer to lumber camp life, he asked what was happening, and he was told there was going to be a fight. The bunkhouse had the usual pole bunks, both upper and lower, with straw mattresses. Some were near the stove, others were a considerable distance away. There was a dispute going on over who would sleep where. One chap had put his pack on a bunk near the stove, and he was arguing with a big raw-boned man named Bob Forbes. Forbes had thrown his pack on the same bunk and had thrown the other chap's pack on a bunk near the cold end of the building. Forbes cut two X's in a pole on the bunk and said if anyone wanted it he would have to fight him for it. They argued for some time, and Gillespie didn't go in the camp, he didn't like what he was hearing. Then these two emerged and the men formed a ring around them in the snow. They fought it out with bare knuckles. They were both sizable fellows, but it only took a few moments for Forbes to overcome the other fellow. The boys told Gillespie afterwards that Forbes was called the "Bull of the Bush." He was a young, husky man and when he wanted something he got it.

ERNIE CARLTON (b. 1891)

This step-dancing business is just a kind of a gift, I guess, because I never tried to learn it. It's just in my head, then my feet. It's like a violin player with his fingers. The first getting up before a crowd was when I went out west after the first war. The more I did it, I seemed to get a little better all the time. Of course my old man could step-dance. He died at ninety-two, around Christmas, and that previous fall he step-danced at a dance at Bell Lake.

W.T. LUNDY (b. circa 1910)

There was a whole bunch of us from Dunchurch who went up [to Port Loring] that year to get a job. Work was hard to get. We didn't have any place to go on Sundays, no place much to go. Always a big table-full Sundays. You were away back in there, nobody to bother you — unless you got lousy. We did go out to the beer place in Loring. It wasn't a hotel, it was just in this woman's house. You walked in her house and there were tables there just the same as a beer parlour. When she lived in Dunchurch she used to make homebrew, liquor out of potatoes. She could only sell Sudbury beer. The company came to her and told her if she'd sell their beer and not anything else, that they would pay her first fine. She ran that place all that winter and all summer up to fall before [Provincial Police Constable] Finger caught her. The company paid her fine and she quit, she didn't sell any more. I never went to the woman and told her I knew her, but the next time I went out there her husband said, "Are you one of T. Lundy's boys?" I said yes, I was. He said, "When you're going home tonight, when you go out through the door, walk around the corner of the house and there's something there I want you to have." I didn't know what it would be. I had my brother Emerson with me, and I walked around the corner of the house and here was a case of beer. That he'd took from his wife! He just went in the basement and shoved it out the window.

NELSON CLELLAND (b. 1899)

Ever play Hot Hand? You'd get a fellow down in the hat and put his hand out flat and slap him. He had to guess who it was and if he guessed the one that slapped him, the one caught was the one that went down. One Saturday night they were playing Hot Hand and there was a lad that got kicked in the head when he was young. He would take spells, once in a while he'd go kinda off. I knew him from the time he was so high and used to work with him, kind of keep an eye on him. This big fellow he took one of these long shoepacs by the leg and was going to swing it to hit Russell's hand. I was behind him and just snapped it out of his hand and threw it under the bunk. Boy! He went up in the air. "Just come outside!" I figured I was in for a licking, but I says, "All right, if that's the way you feel about it, but I ain't going to see nobody get hit that way." I started for the door and Arch Brear, he grabbed the poker and beat me to the door. He says, "There ain't nobody going out of here!" I didn't mind sticking up for the lad. I'd knew him all my life; somebody like that, you'd be willing to take a licking to save him. That's just the way I felt that night. I figured I was in for a licking, but I was going to do my best — I guess it was the Scotch in me. But Archie Brear he stopped it right there. There used to be quite a bit of bullying like that.

JACK McAULIFFE (b. 1901

In the camps there was usually some guy who could sing the shanty songs — "Breaking the Jam on Garry's Rock " "The Foreman Young Munro." They would sing these songs, then they would write them out for some guy, and he would give them a package of tobacco or a plug of chewing or something. You'd have a little sing song at night. Maybe somebody had a mouth organ. The lights were out at nine o'clock.

I remember writing letters on Sundays for fellows that couldn't read or write. This guy wanted me to write to his girlfriend. It seems to me she was French-Canadian, up in the Massey area. He was lonesome for the sweetheart he left behind, and I put all kinds of love stuff in that he didn't know. Anyway, she never wrote back. He probably gave me a package of tobacco.

Schroeder was pretty up-to-date, although I heard they never made any money. In 1918 there were mainly boys and elderly men in the camps. All the able-bodied men were in the army. I can understand kids in high school today, they're pretty unsettled — "Which direction am I going to go?" When I first went in I used to saw with this Polack, and he used to give me shit. I used to ride the saw. You just had to take it, because as I say, we needed a buck. But it's good for you to be in that state. It doesn't hurt a bit. We've spoiled our grandchildren. We didn't mean to. We didn't want them to have to rough it the way we did. But it's been a mistake.

December 2, 1893: Sam Oldfield and his brother Bill got pretty well pounded by the Montgomery gang and Jim Montgomery was run into the lock-up. — D.F. Macdonald

Lunch in the bush in Lount Township. — FRANK WALDEN

GRANDFATHERS, FATHERS, MOTHERS AND SONS

JIM McINTOSH (b. 1896)

My grandfather hewed timber for J.R. Booth, lumber king of the world, and my father hewed timber for him too. They felled them with axes with forty-inch handles. There were no saws them days. Then they'd take three feet off the butt to square it off. They always figured that piece was cull. It would be if it went into lumber, because it would check. There were these blocks left all over the bush. They'd be out there at daylight and out there at dark, as long as they could see that line on the log. I heard my old grandfather say he had hewed a twenty-seven-log house and a thirty-two-log barn, and he only had one arm. His arm was off, but he could use a scythe, a cradle and a broad axe. He'd stand with his back to the tree, look over his shoulder and chop with one hand. Jim Hazzard, he worked on the river all his life and he only had one arm. The only job he couldn't do was tie his tie. He was just as good as any man with two arms.

HECTOR WYE (b. 1905)

Dad kind of followed McCallum. He worked for him [in Hagerman Township], and I can really remember 1913 because that was the year the barn blew down, the Good Friday windstorm. I don't know where my dad was, but he went in by International Siding and down to Big Wilson Lake. The old loggers used to pretty near go crazy in sleigh-hauling time. There was always the danger of coming break-up and leaving the logs in the bush. They didn't want to get caught with that, so they usually finished in good time, but they weren't that year. Anyway, the barn blew down and he came home the next day, down the Whitestone road with the team and sleigh. After being away working all winter for not much wages, and to come home and find the barn blown down.

The thing was to get the team away. It wasn't exactly the money the team brought in, it was also to get the team fed. They quite often came out in better shape than if they'd stayed at home — used to working and well fed. When Dad was hauling logs at Boakview I'd go down on Saturday night and look after the horses and he'd come home for the weekend. The weekend was just Sunday then, there wasn't even any getting off early on Saturday.

The year of my fifth birthday, the first birthday I remember, Dad had a cutting and skidding job of his own [behind the farm]. Graves and Bigwood I guess had the timber rights, but he was jobbing for McCallum. That must have paid pretty good. My mother said he paid off what was owed on this place, and that was the year he bought the Sands place, that spring. One thing my dad could do was look after bush work. When he was back here, I remember the men walking out past here. I guess it was the easiest way to come out. I remember there were a whole bunch of men, it took them a long time to go by, two or three at a time. We were watching to see if we could pick out which one was Dad. Sometimes men going back in to work for McCallum would be here for supper. They'd be on their way in or out and Dad would have them in to have supper or maybe stay overnight. These Polacks, we didn't know whether they were dangerous or not. We used to be scared to open our mouths. Whether they were Poles or not, I don't know. They used to call all those European people who came over "Polacks." Everybody complained amout immigrants coming in. They thought McCallum favoured the Polacks. I guess maybe they'd do what he wanted when other men wouldn't.

MEL CAMERON (b. 1892)

My dad took me pretty near every place he went. When I was thirteen I went down with him when he worked for McCallum around Dunchurch. McCallum was a short fellow, a comical guy and pretty talkative sort of man. He had a pretty nice driving team. He had a buffalo robe and fur coat. He was the greatest walker. There was nobody could cover ground better than McCallum himself.

A curly-haired guy, Charlie Piatte, was foreman at Number One Camp, alongside a marsh. They wanted my dad to go to the other camp that Jack Campbell was running, so he took me and went over there. A big stout fellow they called Dollar — his real name was Madigan — had a big pair of oxen there, cleaning off the main road, falling trees across it and moving logs off. I was helping him brush out. Then when they started to haul we came back to Number One. My dad was a pretty good hooksman and they put him top-loading. There was a team of oxen there too, for breaking main roads. An ox can go where a horse can't go at all — break roads, snowplow. But an ox can't stand on ice, can't pull his own weight on ice. They put shoes on them for on the ice road. On the start of the sleigh haul Charlie Piatte put me cutting tankholes. They had to haul over a grade, and I don't know why — I think he done it more as a joke — he told me to get these oxen yoked up and bring them out, stay at this grade and tow the teams when they came up with the logs. Well, me, a kid, and a big pair of oxen. Hah! They got so darn lazy they'd hardly tighten the chain for me. They'd take a look at me, and I was too small. Then Charlie'd come and take a hold of them. A blacksnake whip. He'd tune 'em up and they'd work pretty good for a little while. But I drove quite a while there and got them over the grade.

It was pretty near three mile out, close to Shawanaga Lake. We used to walk that early in the morning. It was pretty thick pine and they'd have to fall trees in the dark. They gave us breakfast in the middle of the night. McCallum said if you slept after three o'clock you'd get bed sores.

NORMAN CAMERON (b. 1894)

My dad [Allan Cameron] ran away from home when he was fourteen years old. He took the Champlain Trail and made her all the way across to Montreal. He worked his way, wherever he could get a day's work from anybody. He went all the way to Montreal and made friends with a Frenchman, Paul Ouelette. His father was a blacksmith and he had learned the young lad to blacksmith too. How Dad happened to come back here from Montreal, Paul was telling him, "You should see my girl. I'm going to get a team and take her to church. You'll see us going down the street." So Paul gets a team Sunday morning, drives down, picks up his girl, and she's all dressed up with a big hat and flowing skirts, a really nice-looking girl. She had an umbrella, and when they got out in front of where Dad was staying it started to rain and she stuck up her umbrella. The horses got scared of the umbrella and started to run. They upset the buggy and smashed it and harness all to pieces. Paul and Dad caught the horses way down the street. Paul had no money to pay the damages, and him and Dad struck off to Nosbonsing Lake to take a job they'd heard about. Dad fired on the little railroad where they hauled logs out of Nosbonsing Lake and dumped them in [Callander?] Bay. When he was twenty-one he went driving team for a fellow named Joe Driver, the first postmaster in this part of the country. They sent him out to my grandfather Whitehead's to get potatoes — he sold potatoes to the lumber camps. My mother was just fourteen going on fifteen, and she was down in the cellar picking the sprouts off these potatoes. He rapped on the door and she said, "Come in." You know when you come indoors on a frosty morning you don't see anything. Well, he fell into the cellarway. That's where he first met my mother, in the cellar.

My dad and mother cooked for Albert McCallum in back of Dunchurch. That'd be 1898. Mother was the main cook and Dad was the cookee, but later Dad began to cook himself. The camp was in a gulley between two hills. Everything was brand new. Where they dug the well was six feet of solid blue clay, and they plastered the camps with the clay they dug out of the well. They put moss on the inside and that clay plaster on the outside. McCallum bought a cow and brought it in to the camp. No canned milk them days. He told Dad, "Now you milk that cow and you'll have milk for porridge." So Dad said he'd milk it. He'd never milked a cow in his life. He started to milk her by the well one night, and by the time he got the cow milked he had dragged her pretty near over to the stable, pulling on her. He pulled too hard and she kept stepping over. He thought he had to pull instead of squeeze.

I'd say there'd be about forty men in camp. They had a cookery and the office, then they had the stable and a hayshed and granary at the end of the stable, all built of long pine timber. The cookery would be about forty feet long, maybe forty-five, and twenty-five or thirty wide. They had two big long tables. And in a corner they had a bedroom partitioned off for my dad and mother. They had a ceiling over it, and my brother and I slept on a bunk bed on top of that.

McCallum was a live wire, on the go all the time, hardly ever sat still, only long enough to eat. In one day and gone the next. He had his cutter and a big buffalo robe he'd sit on. He had a team of collies, and they'd take him from one place to another. He had no line on them, they'd just go any place he told them. We had a sister only two years old, and he'd take us kids all over with him. My brother and I used to play logging. McCallum got us a little saw — I have it up in the woodshed somewhere yet — one of them saddleback saws. And we'd saw limbs off and pile them up on a little sleigh that we made, then hitch the dogs on.

We stayed all winter [and the next summer] we lived at McCallum's house up on the hill in Dunchurch. That was the first place I heard bagpipes. Billy Buchanan had the barbershop, and when there was nobody getting a haircut, he'd get the bagpipes and start up and down the length of the barbershop. I used to run around with "Boots" Robertson, who belonged to the storekeeper. He was a big boy. He had feet bigger than any man in the country, and he wore boots two sizes too big for his feet. We'd hear the bagpipes and away we'd go, we'd run and stand at the door. And I always wondered, why does he keep walking? He'd walk up to the door and turn around and go back the other way and back again. First thing you know he'd have somebody there to get their hair cut. They'd go to hear the bagpiper, and while they were there they'd have a haircut or get a shave. I have always liked bagpipes.

Wherever Dad worked he got the nearest house for us to live. When I was two years old he hired an Indian to bring Mother and I and Mel to Wahnapitae, to stay with a family named Brown. We went in a birchbark canoe down Lake Nipissing, down the French and back up the Wahnapitae. I can remember that just as well as if it was today.

BOB GIBSON (b. circa 1892)

My father [Hugh Gibson] and Uncle Frank and Uncle Billy came from Ireland to Toronto and worked for a year for a farmer where the Exhibition Grounds are now. Then they came to Owen Sound, across by boat to Parry Sound and up to Croft Township, where they built a little shack and squatted. There was no one else in that area at the time. After people came and started to log he worked out. The first logs were hewed in the bush. It was boat timber, to go to England. If a tree was good enough — sixty or seventy feet without any knots or any punk in the butt — they'd take that tree out. If it wasn't, they just left that tree and got one that was. They slooped it to the Magnetawan River. Johnston and Beveridge was one company I heard my father speak of. He worked out for only a month. He got $8 for it, and he figured he'd never make a go of it like that, so he went down to Parry Sound and got stuff, toted up through here, and moved settlers in. When they got the road through from Parry Sound to Gravenhurst — as far as the train came in the early times — he got horses then, and he'd go to Gravenhurst and buy flour, lard and tea, tote it up and sell it, and go back for more.

He bought cattle for the lumber camp in at Island Lake [Wilson Township], bought them in the farming country below Gravenhurst and drove them on foot up through Falkenburg, Rosseau and Parry Sound, in through Whitestone and up around the far end of Deer Lake. He killed them at Farm Creek, took the hides back when he went for a load of stuff, shipped them at Gravenhurst and bought more provisions. Cattle weren't too dear then.

He'd have a couple of men with him. One night down where the Holt Timber Company later built the railway down to Deer Lake, the cattle were getting off in the brush on the side of the road, hungry, browsing. So he built a fire and fed the horses and blanketed them, and took five or six logs out of a pole bridge at Maple Creek where it comes into Deer Lake so the cattle wouldn't go back. Stayed there till daylight, put the poles back in the bridge and took his cattle and went through.

JIM LUDGATE (b. circa 1900)

My father's father, Peter Ludgate, was a lumberman working out of Peterborough. My dad [James Ludgate Sr.] was the eldest of the family, and he ran away from school and headed for Parry Sound when he was maybe sixteen. He worked in a mill. While he was still quite young he got on as assistant scaler. He got $16 a month and board for that. But the scalers wouldn't start work as early as everybody else in camp, so he got $2 extra a month for cleaning out the horse stables. And that was his start. He worked around the Seguin River waters mostly. He contracted to bring drives down for Peter's and I guess maybe the Parry Sound. We lived, as far back as I can remember, on River Street. There was a big stable out behind, about where the CNR shops were. I can remember punts and pointers and equipment, hand-operated winches, that type of thing, stored in the stable.

Dad was also walking boss for Peter's. A walking boss was the fellow actually on the job for whatever lumber camps they might have, head foreman of the camp foremen. There might be a general manager, depending on the size of the firm. Dad and Mother lived at Seguin Falls, that was the centre of their limits. Alvin Peter's father was a real old German. Dad had a very high regard for him, thought he was the finest man he'd ever worked with. He was there till the old man died. Dad and Alvin could get along, but I think Dad pretty well carried him. Alvin squandered his money.

FRED COURVOISIER (b. 1905)

My grandfather came to Poverty Bay in 1877 or 1878. He looked over the country and there was this big beaver meadow back here. He thought this would be a good place to stay; he could cut some hay. He had fetched cattle over from Switzerland. At that time there were two lumber camps north of Poverty Bay, one called Campbell and they other they called the Tame Meadow. They kept men working in there the year round, and they kept horses the year round. That's the reason they cleaned up the place they called the Tame Meadow, to grow tame hay to feed the horses. They had a storehouse on Poverty Bay and a cadgeroad back to the Tame Meadow.

ALEX GALIPEAU (b. circa 1908)

My grandfather worked in square timber. I turned the crank on his grindstone to sharpen his broad axe. Only sharpen on the one side, just one bevel. It had to be just so. Maybe an hour and a half or two hours. Holy Jeez, I was tired. I saw him doing the work when I was eight or nine years old. It was for Hettler Lumber Company on Georgian Bay. It was mostly red pine. They rafted it at the mouth of the Sturgeon River and took it across to Chaudiere Falls. They put some bedding down before they cut the tree down, to carry it. First of all the branches would be cut off and there would be a strip [of bark peeled off] with an axe, then it would be marked with a line with black charcoal on it. There would always be some green alders burning in the fire. There was a man standing up on the log to block the tree, notch it here and then notch it there, pretty near to that [line], then split that off. They'd be around eighteen or twenty inches long, the blocks. That would be on the big end; when they got to the smaller end they'd be longer. He had a four-pound axe, a Black Diamond head they used to call them, with a handle thirty-eight or forty inches long. Them old fellows didn't like to play with light stuff. When he got to the small end he would start back at the butt again. At the butt you've got deeper blocking, maybe sometimes eight inches. Grandpa would be maybe halfway up the side by the time the man was half finished blocking. Grandpa would catch up to him.

NELSON CLELLAND (b. 1899)

My father swamped tanbark with an ox. They had a sloop affair with two long poles that went back and trailed, and stakes in them. That let them go around trees and swing backward and forward. They said they could pull a four-by-eight-foot cord of bark with that one ox. They'd take five cord of it on a sleigh to Ahmic Harbour, from away back in here. The load would be sixteen or eighteen feet long and two piles eight feet wide. Just imagine that behind a team of horses.

Jack Hosick, I ain't sure whether it was his dad or grandad, he was working up here and he said, "I think I'll take a little run down home." His home was down at Gravenhurst and he left [Dunchurch] running down there on foot to see his wife. Out from Parry Sound somewhere he caught up with an old fellow and the fellow says, "Have a ride?" So he got on and rode for a while. Then he says, "I've got my legs straightened out now, I guess I can move on." Hopped off and away he goes, on ahead. Just imagine, start on foot from here to Gravenhurst. You know, I think in one way they were better times than now. Everybody was working and more content. They didn't have time to get into trouble. They were too busy making a living. Nowadays, look at the crime there is all over the country. Them times, you never heard tell of it.

Bridge where the Thirty Mile Road crosses the Pickerel River near Dollar's Dam. Another version of George Brunne's Wild Wagon story has a drunken teamster whipping his team out onto this bridge on a dark night when part of it had been carried off in a flood, and man and horses disappeared without trace. — GEORGE KNIGHT

Neil Conroy, handyman, and Jim Wilson, blacksmith, at a Schroeder Mills camp in Wilson Township, circa 1913. A stock of blacksmith's steel leans against the wall.
— GEORGE KNIGHT

Sunday in camp was a time for homemade amusement. This mock hanging was staged for an itinerant photographer. — EVERETT KIRTON

Bill McKelvey (l.) and David Sword (r.) were widely known woods foremen of the 1910 era.
— GEORGE KNIGHT

MORE BUSH CHARACTERS

JIM McINTOSH (b. 1896)

The Grand Trunk met J.R. Booth and wanted to buy his railroad. They asked him what he would take and how many years he would give the Grand Trunk to pay for it. The knees were out of his pants and the elbows were out of his shirt. He looked up, smiled and said, "What would *you* take for all your railroads in Canada, and I'll pay you cash!"

BERT LITTLE (b. circa 1885)

There were some hard cases. These Dutch people named Sauder were settlers on Shawanaga Lake, and the husband went away to the lumber camp, leaving her alone and expecting a child. By the time he got his first pay and came home, she had had the child. She was completely out of food, and she had gone out and gathered dead beech leaves and boiled them for something to eat.

BURLEY HARRIS (b. 1883)

There'd be a peddler come into the camp with a box of jewellery. He'd make quite a bit. Just before Christmas he'd come along, the lads would buy something for someone at home, and first thing you know he'd make a pretty good pull out of it. I paid $28 for a watch and chain.

NORMAN CAMERON (b. 1894)

The following is Norman Cameron's account of an armed robbery which occurred around 1922 at the Schroeder Mills and Timber Company office at Pakesley. James Ludgate Sr. was woods manager and George E. Knight the clerk. Knight took many of the photographs reproduced in this book.

It was getting dark at night, and George went to bed. Ludgate was sitting out. It was summertime and flies were awful bad. But Ludgate was a tough old sucker, he could stand a lot of flies. George just laid down on the bed and he heard these fellows talking real rough to Ludgate. They had come and held him up. They had worked [for Schroeder] at some time and knew where everything was. In comes one fellow to make George open the bank. George always kept a .38 automatic revolver under his pillow just in case somebody did come to hold up the office. George heard the man open the door, and he reached under the pillow and grabbed this revolver. George's door was open. Back down the hallway there was another room that Ludgate slept in. The office was at the end. George got his revolver and was just getting out of bed when he saw the fellow jump past his door, looking for him. The man had a revolver in his hand, and George let go and hit him in the hand, cut two fingers pretty near off, hit the butt off the revolver and knocked it out of his hand onto the floor. The fellow picked it up with his other hand and ran into the back room. He was afraid to come out of that room and George was afraid to come out of his. There they were, one in each room with a revolver. Ludgate was on the verandah and the other lad was holding a .32 Special under his nose.

Ludgate yelled, "Come on out, George, you might as well let them have it. There's no use killing anybody for all the money there is around here." So George came out, and they hit him over the head and knocked him out. When he came to they made him unlock the safe. They took everything that was any good — the gold watches and the rings [employees] had stored there, and what cash there was. They took George and away they went walking up the railroad track. George was in his bare feet in the cinders on the track, at night in mosquito time, and just a thin little white shirt to sleep in.

They walked pretty near to Pakeshkag Creek, about three miles, and there was a spring there on the left side of the railroad. It's steep down the bank to the spring. The one fellow took the other one down to wash, and he tore a chunk out of George's shirt to bandage his hand with. They said they were going to shoot George. It was good and dark, and George took for the bush on the other side of the track. When they got up on the railroad he was out of sight. He kept running and hiding, running and hiding, but they only went after him apiece and then went back. They thought he'd die before he got out of there anyway, him bare naked and all them flies, his feet bleeding and played clean out from walking three miles in them cinders. But George sat there and didn't move till ten o'clock next morning. Men and police were hunting for him along the track on handcars. They found George, but they didn't find the robbers then. They went up to Sudbury. There was a farmer living just out of Sudbury and they went in to rob his hen roost. The farmer heard the hens, thought it was a fox or something, and got the shotgun. And by golly he held up the one lad. The other lad came back down here to do some more robbing, came in to Lost Channel. He must have come up the French River and up Dollar's Lake. The police were trailing him all over, and they found him. He seen them coming, run down and got in his boat, took the butt of his rifle and pushed the boat out into the bay. The wind kept drifting him out. He just stood there with his gun on them, and he says, "The first one picks up your gun, I'll down you." He drifted out that far they couldn't shoot him with a revolver, then he put his gun down and took his oars and rowed away. But they got him later and he got five years in jail.

HENRY NORTH (b. circa 1900)   LEVI NORTH (b. 1898)

There's a grave between Deer Lake and Bolger station on Holt's railway. No man's grave. I saw the cross when I cadged that winter. There was no name on it and I never knew who it was. I figure he was one of Holt's men. I guess there were quite a few [river-drivers] didn't make the [Byng] Inlet. Once old Dollar [James Madigan] came to a grave down at what they call The Graves. Old Jim was looking at the graves and there was a pair of boots and a pipe there. He said, "Jeez, I wonder if he wants them anymore. I'll just ask the Lord and see." He looked up, then he said, "No, he didn't say nothing, so I guess I can have them."

DON MACFIE (b. 1891)

"Scaler" McPhee stopped at the Wildcat Hotel and a girl working there said, "Mr. McPhee, how would you like to scale me?" He said, "I see a pair of punk knots, a gum seam and a big turned butt."

JIM McARTHUR (b. 1886)

Joe Simpson was a jobber for Graves and Bigwood. He came to [South Magnetawan] about 1904 and started lumbering there. He wasn't married. His people were English and he took the London *Times*. I was walking the track up above Burton and he was coming down to see me, because he had nobody else to talk to. He says, "Come on back." So I went back with him. He lived in a log cabin maybe twenty-four feet long, beside the track, a place for his men. We had a talk, then he brought out the London *Times*. He said, "Look at that, Jim. That's the woman for me! You see what she done?" His grammar wasn't proper. "You see what she done?" he says. "Just look at this!" Two burglars had broken into this cottage and this woman had them by the scruff of their necks. There was a picture of her in the London *Times*, on the front page. This woman weighed 180 pounds and was six foot tall. He laid the paper down and says to me, "Help me write a letter." So I wrote a letter for him, helped him, a love letter. I should have been a lovelorn writer! Hah! I put in some nice words and one thing and another. He put in that letter the most lies! He was a lumberman. He had $3,000 in the bank. (I don't think he had $50.) So many men working for him. He was a boss, you see. I put it down. I thought, it won't matter anyway. Anybody that would fall for that is crazy.

But he sent the letter, and it got to London and they published it. And his picture, he sent his picture with an old Methodist hat, the big black hat, and clean shaven, a good-looking fellow. There was his picture, you'd have thought he was a millionaire from the write-up, this big lumberman with his big estate and all this. And she fell for him.

Another letter came in the London *Times* — not on the front page this time — that Miss

So-and-So was engaged to Mr. Joseph Simpson of South Magnetawan, Canada, and they were going to be married — nuptials, in those days they were nuptials, I remember that. I have the paper yet, he gave it to me and I kept it. Gee, he was going around there and you never seen anything like it. What's he gonna do? He's got to get to London. He says, "You know, I haven't got a heck of a lot of money." And he had told her he had $3,000 in the bank. Well, you should have seen him when he got dressed up. I don't know how he did it; some of the dealers in Parry Sound must be broke yet. Two pairs of boots, two hats, three suits of clothes. The stuff he took, three big grips. "Boys, I'm going in style," he said. "I might meet the Queen." And away he goes to London. We all wished him well. I forget the fellow he left as boss there, he looked after things. And away he went.

He got over there, and they came back and were all right at the start. While he was away this boss and the men had built a new house by the railroad, a log house with five or six rooms. She had a piano and a gramophone, one of these cylinder ones, an Edison. Oh, beautiful carpet on the floor, some new furniture and the kitchen was all up-to-date, new stove for her and everything. He must have had a little money more than I thought he had. But I guess she got kind of lonesome and fed up, nobody else around. One day about two or three months after he was married, I saw him coming down to the river from his house. He had just bought a new pointer, thirty foot with a Beaver inboard motor in it. He had that for hauling logs. I could see him, and tin cans and frying pans after him. Swearing! That woman could swear more than a lumberjack. What she called him! Oh, it was terrible. Joe shoved the pointer out, jumped in and away he went down the river to his other camp. That went on for about two years, then she had a baby, a girl, and then a boy. Well, they seemed to get on better after that, and he was cutting more and making more money. But anyways, they had another flare-up and she left him and went to Toronto. Some relative of hers came over and they came up to take the children. Joe wouldn't give them up, but they got one. He took the girl, and he went to Alaska. She was so small he put her in a packsack. And do you know he became the mayor of Fairbanks, or one of those towns. And the girl, she grew up, and I have a letter yet that she wrote me. She was fifteen or sixteen. "Is there any chance of me coming back to South Magnetawan?" Where she was born, you see. It was still in her, what he told her about it. I wrote and told her, "The place is gone. People are here, it isn't quiet like it used to be. You wouldn't like it." So I never heard from her again. He wrote to me several times. He said, "I'm doing well. Come on out here, you'll make a fortune in no time!" He never married again.

That was Joe. Joe Simpson. He got in wrong with Graves and Bigwood. They tried to claim he was in their debt two or three thousand dollars. And he had built a big place at the end of the river, South Mag Camp, to take in tourists. But she got out and he couldn't do it [alone]. He lost that and they took anything else that was in his name, seized everything. I thought it was wrong, because I found Joe Simpson an honest man, one of the old-timers raised in old country style.

MARSHALL DOBSON (b. 1891)
"John Dollar" [James Madigan] was a stout man and a big eater. He would eat so much meat and potatoes that one day the pie was all eaten up before he got to it. So next suppertime he stuck a ball of horse manure in his pocket, and when he went in the cookery and sat down at the table he drew a big X across the top of a pie with it. It was still there when he wanted it.

November 14, 1877: Ross was ordered to the Central Prison for stealing tools on the PSL [Parry Sound Lumber] Coy. — *D.F. Macdonald*

Shown above is a footbridge placed over the Seguin River's Serpent Rapids by river drivers. Below is the Magnetawan River cadgeroad where it spanned The Graves Rapids. — FRED LEUSHNER

— JIM McARTHUR

LOST TIME, LOST FINGERS

JIM McINTOSH (b. 1896)

Go into a camp in September with a hundred men and there wouldn't be a man you'd have to take out. When we used to go to camp in 1916, 1917, they'd take in five gallons of turpentine and five gallons of coal oil. If you got a sore throat you took off your woollen sock, turned it inside out, put the sole right to your throat and put a safety pin behind. In the morning you had no sore throat. That's a sure, sure remedy, a woollen sock. And a man got a cold, he'd go to the cook and get a handful of brown sugar, pour lots of coal oil on it, take a mouthful of that, and next thing he knew he had no cold. I've got it right here. I don't mind drinking coal oil, turpentine.

I lost that finger at four o'clock in the morning, when I was running camp back here. The men shouldn't have went out at all, but I said to them, "If you get an early start for two mornings and finish at noon, I'll allow you the full day, and you'll only have two more days to work." They all hit for the bush, and I went right out with the men. By the time I got there they'd sawed a big birch and the saw had bound — they were trying to saw it the wrong way. So I says, "OK, boys, I'll fix that for you." They had two thin iron wedges, and I was going to make a wooden wedge to drive in to tip the tree over for them. I took my axe, knocked a piece out of the tree and set it on top of a stump. It was dark and when I went to make the wedge the darn thing slipped off the stump and my axe came down and cut the thumb and that finger. The finger fell off. I just picked it up and put it in my pocket, and they felled the tree. It never hurt me one bit and I never saw a doctor. I came to the camp and old Johnny Wylie put two stitches across with an old buckskin needle, pulled it together.

GOWAN GORDON (b. 1909)

I broke my leg sawing for Sticklands at Port Loring the winter I was eighteen. We cut a hemlock and we didn't notice it had a bind. I was down on my knee sawing left-handed, with my right leg ahead, and when we went through, the butt flew over and went down on my leg and broke it. Some of the lads carried me a mile out to Toad Lake on their backs. It was the second or third day I'd worked there.

JIM McAMMOND (b. 1903)

My father was working at Blackstone out from Parry Sound. He was going down a sand hill, and the log he was sitting on went out between the horses and he went under the sleigh. There was just this one log thrown loose on top of the load. He went under the sleigh, and it went over one leg and threw one hip out and broke a leg. His hip never was put back in the socket, and he was crippled.

WILLIAM McKEOWN (b. 1885)

One time I cut myself with the axe in the camp. It was cut down below the knee, with the corner of the axe. The lads did everything they could, put flour on and tied it up, but it still kept oozing and bled all night. As luck happened there was an Indian going around selling herbs, medicine. He had a good team and a buggy. Next morning he was coming out, and he drove me right home.

NORMAN CAMERON (b. 1894)

I started to fire-range for the government when I was seventeen years old. That's after the winter we cut ties and the tree fell on me — smashed two ribs and one went into my lung. That's what kept me out of the war, this bad lung. Them days you didn't get to a hospital. The doctor just put a thing on to hold the rib out. I took an abscess in the lung and they gave me three hours to live. That's a long time ago and I'm going yet.

We were cutting ties, a whole bunch of us — Fred Hampel, Frank Dellandrea and Mathew, Sank Whitehead, my brother and I — just finishing up. We'd made our contract, but when the scaler came you had to have 10,000 ties over and above the culls. Them days they culled everything they could on you, because they got it for nothing. We got seven cents for a cull, thirteen cents for a number three, seventeen cents for a number two and twenty-one cents for a number one tie, hewed hemlock, cut eight feet long and dumped in the lake. Well, this little hemlock lodged against another one. We sat down and had dinner, and after we ate our dinner somebody said, "Let's push that tree off there." So they all run around on the opposite side to push the tree this way — all it needed was a little push. I was the last one getting there and I just went up to this side and pulled on it. I intended to step back, but my heel was behind a little twig that was in the snow, and I fell backwards and the tree rolled down on me. I was in this little hollow. It just squeezed enough to break the ribs on one side. That put me out of the race.

ROY COCHRAN (b. 1905)
There was a doctor who came into the camp once or twice a month. Everybody paid a dollar. Dr. Denholm was into our camp and Doog Campbell's camps, and I don't know how many others. I had the quinsy once when he came in and he made me go out with him. But I found a cure of my own for quinsy after that. Your tonsils would swell up, swell up enough to choke you. I ran into this horse Absorbine, and if I was getting quinsy I'd feel it up behind my ears and I'd rub that horse Absorbine on the outside three or four times real good, and it would go. I haven't had it for long enough.

Old Sam Saad, he went underneath a tree that was hung up, to let it down, and it came down and struck him. I thought the man would never live. It drove his boots four or five inches in the ground and the axe came up and drove a three-cornered hole under his chin. We thought he was going to bleed to death before we got him out. He lived but he was crippled after. The only way you had to get them out was with horses, there was no other way. When you came out to a railway, no matter what train came along you'd flag it down if a man was hurt.

DICK BREAR (b. circa 1900)
That burnt timber [on the George Holt Company limit in McKenzie Township] stayed there quite a few years. I think that fire started around Delta Marsh. Somebody burnt that stuff to get them started working. It burnt from [Holt's] railroad right through to the Pickerel River. At the time of the bad flu I was in Camp Nine. They quarantined us in camp, none of us could go out. I think it was three or four weeks. Uncle Joe Ward was supposed to be the fellow that watched that nobody went out and nobody came in. He didn't work in the bush and they picked him out as a kind of policeman. That's where the old doctor gave me up, I was so bad. The old handyman looked after me just like a nurse. When I got out of bed and started walking, I had to learn all over again. I had to follow the bunks around wherever I went; I couldn't start off and walk or down I'd go. My brother Percy and I went out home one Sunday after I got so I could walk, to see how things were. There was about a foot of snow and my eyes were so bad I had to stop and look for his tracks. That was a bad flu, there was so goddamn many died. It got so bad they wouldn't go to funerals. Scared.

Up at Schroeder's, the main road gang was tramping roads and this newcomer fell in the bloody tankhole. In place of the buck beaver sending somebody with him into the camp, they let him go alone. They were breaking roads here, there and everywhere all through the bush for the sleigh haul, and he got lost. He got on the wrong road and didn't make it to camp till suppertime. A rap came to the bunkhouse door, and when I got there this fellow was fooling with the latch. He was so cold he couldn't get the latch open. When the door opened he went right in on his face. They pulled his clothes off — used a knife on his feetwear — threw him on the bed and went at him with white linament. I betcha they worked on him half an hour before they got a word out of him. The buck beaver got hell; he should have sent somebody in with him, him wet to the neck.

October 22, 1918: I stayed at Kaufman's [lumber camp] in Shawanaga. They are all sick with La Grippe. October 23: The epidemic is on the increase. October 26: The flu is making blanks all round. — D.F. Macdonald

98

COMPANIES AND BOSSES

BERNARD MOULTON (b. 1894)

Dick Cooper ran a camp for Albert McCallum in from Maple Island, and I worked there. I don't know if it was a camp or a dancing school. Every night after supper they were all dancing in the bunkhouse — a bunch of young fellows, and Cooper would be among them just like a big kid.

Albert McCallum was a short, stout man. He smoked cigars and the biggest part of the time he wore shoepacs. He was a good-hearted fellow. He thought Polacks were God's chosen people. But long hours. One Sunday morning he was sitting on a set of sleighs and Davey Black says, "Say, Mac, I thought this camp was log." McCallum said, "Did you work in here all this time and didn't know it was frame?" And Davey said, "It's the first time I saw it in daylight!"

I was driving team at Wilson Lake and I was second team. McCallum came into camp about half past four in the morning and said, "Are you not out of here yet?" I said, "No, the lead teamster has just gone out." "Gosh," he said, "I used to be on my way to school this time of the morning!" Half past four in the morning.

McCallum had a foreman [on the river drive] but sometimes, if there were logs jammed on the shore, he would get in the punt and go out and help shove them out. I remember one time on the Shawanaga River a fellow named Tom Shannon came in and McCallum asked him if he could drive, and he said, "No." So McCallum took him in the punt with him. When it came noon McCallum said, "You follow me, we're going ashore for lunch." Every log McCallum jumped on, he jumped on it too. He had McCallum halfway to the knees all the time. Why, that man could ride a bubble. He was just playing a joke on McCallum — paddled him around among the logs all morning, then come to find out Tom was the best driver he had.

It was the seventeenth of March when the races were on the ice at Ahmic Harbour, at the last of the sleigh haul. All the teamsters made out they'd go to the races. They had quite a few logs to get out yet, so Frank McCallum [Albert's son] said, "Any man goes to the races gets fired." So every teamster went to the office that day, got their time and went to the races. [Albert] McCallum went out the next day and hired them all back.

McCallum made money in Number One Camp. That was his very best. He seemed to get the drive out every year. When he came out of that Hagerman limit, they tell me he had $75,000 in the bank and twenty-four teams of horses, plus harness, sleighs, cookery, rigging, everything. And died broke. It was Holt's that sunk him. He was supposed to put the drive out, and he put it over the Burnt Chutes and let the water all away, and that's where it stayed. The Holt Company came after him for it. They had a lawsuit in Parry Sound. They took it to Toronto and they took it to England. Of course he had to pay a lawyer all this time.

DAN CAMPBELL (b. circa 1880)

I worked for McCallum out of Number One Camp one year. We started in July. McCallum came here to Waubamik in a team and buggy and got me to go up. "Going up to start to skid," he said. They had an awful fire. You could hardly drive through Sunnyslope clearing with a horse and buggy. You couldn't see. McCallum got an awful lot of logs burnt. People chopping fallows and burning brush sometimes weren't careful and it would get away and burn a lot of timber.

We got the horses into camp and fed them up for a while, then I cadged for a month. That was quite a good team I had; they had run all summer on the grass and were fat. The road was rough. To let the horses rest, they had what they called a "dog" fastened where the two hounds were on the back of the wagon. You had a bit of rope to keep it off the ground so it wouldn't wear off, then going uphill when the team got tired the "dog" buried in the ground and let them rest. I was cadging stuff from Ahmic Harbour, and after a week or so they sent me out to your grandfather's, Frank Macfie, to get a load of oats. That's where I saw your dad first. Roy would have been about eight or nine years old, and I remember him saying he'd like to be up on that pole seat on the wagon.

James Ludgate Sr. (l.), woods manager for Schroeder Mills and Timber Company, on his rounds.
— JIM LUDGATE

Albert McCallum, a renowned Parry Sound District logging contractor.

George Knight at the Schroeder Mills office at Pakesley.
— GEORGE KNIGHT

McCallum had five or six gangs cutting in the bush, then they started to skid. We used to skid 350 and 400 logs a day, the timber was that thick, but it was just like stovepipes. It was all bald rocks with a bit of moss on them. I don't know why they cut trees as small as they did. They took them out four inches at the top end.

Oh, it's wonderful the way they changed things. The first canthooks [were simply hung from] a band around the stock. There weren't any decking lines for years. When they first had the line, the fellows would as soon [continue to] use the spike skids. They'd get the chain too much to one end or the other and the logs would go sideways. But as men got used to the decking line, they were awfully good. Of course the jammer made it better than that again. They picked the logs up and set them down on the sleigh, saved a lot of work. Dave Wye was a great man with a canthook, and Joe Farley was a good man too. When Dave was top-loading he'd catch the pup when it would come unhooked, catch it up in the air with his canthook and fetch it back down. That was the big trick, to learn how to catch this pup. Some used to catch it with their hands, but Dave Wye would catch it with his hook.

WALTER SCOTT (b. 1893)

The best days this country ever saw were the old logging days. People didn't get very much for their work, but there was always work. I was born up around Seguin Falls, and last time I was up there, there were only two families living in that part of the country. At one time every hundred acres up both sides of the road, people lived there. But they never got nothing much out of the farming. It was just a place to put their spare time. Maybe keep a couple of cows, that's about all the farming they did.

I used to job. I logged for Ludgate and Thompson, and I logged for Graves and Bigwood back in from Pointe au Baril. There was a little lake that didn't have no name and I built my camps alongside it. After I logged there they called it Scott's Lake; I've seen it on the map.

Graves and Bigwood was a good company to job for. They'd advance you lots of money, and they never asked you any questions — at least they didn't of me — and they'd give you anything they could supply you with to help you out. I owned about half the rigging. All I had of the company's was some sleighs and the patent plow. I had my own tank and quite a few sleighs. I had enough sleighs of my own for a small job, but when there was a bigger job I'd get whatever spare ones I wanted from the company. They had quite a few patent plows and they rented them out to you if you wanted one. I guess I might have had $1,500 or $2,000 in equipment — sleighs, tanks, snowplows, saws, axes, wedges, and all that. After I got going a month or six weeks, the company would advance me money. If you were figuring on taking out a million and a half feet of lumber, you wouldn't have any less than sixty men. The logs would run about sixteen to twenty to the thousand [board feet]. You'd have eight or ten teams skidding and maybe eight cutting gangs. A couple less cutting gangs than you'd have skidding teams because when you got a long haul you'd put in an extra team.

You'd see in the paper this company was going to log and they'd want to get contractors to take it out. You'd go and look at the timber, see what price they offered and what you thought you could log it for — whether you could make money or would go in the hole. You'd have to look the whole thing over before you could give them your price on it. Look up where your roads were going, and figure out what it was going to cost to build roads and one thing and another. The company would give you a price in the first place, but sometimes they'd maybe pay more. One time I was logging for Herb Thompson and he gave me a price on some timber he wanted to log, so I went and looked at it, and I said, "I can't log it for that; I'll have to have more money." He said, "I don't think I can pay you more." I said, "That's fine, it's out with me." So he logged himself that winter and he said to me next spring, "I wish I'd given you your price, because it cost us twice as much as you offered to take it out for." It was a bad winter for snow, it was about four feet deep, and it cost as much to handle the snow as it did the logs.

Sometimes you'd get a contract for two years; it just depended on how much timber they had in that location. I used to make around $2,000 [in a winter]. I never lost — only one

Loading the Graves, Bigwood Company's logs on Canadian Pacific Railway flatcars with a jammer near Pointe au Baril. Jack Campbell is at the top of the jammer, while the top loader poses on a bundle of logs held suspended by the horses. — JACK CAMPBELL

A train load of Graves, Bigwood logs leaving the Pointe au Baril area for Byng Inlet.
— JACK CAMPBELL

winter when the snow was deep and the ice was poor. I had five teams of horses of my own, and I worked all winter and fed them, and I only made $300. That winter Doog Campbell lost $10,000 and Jim Lawson lost everything. The company even took his horses and rigging when he finished in the spring, he was so far in the hole. When the snow gets four or five feet deep, it costs an awful lot to handle it. You've got the shovel it off the skidways, and lots of roads, you had to shovel them that winter. The ice was so poor I had to put men snowshoeing to get it frozen enough to get a dump, to get drawing on the lake.

JIM McINTOSH (b. 1896)

I ran camp with 122 men for the Standard Chemical. That was the happiest time of my life. Everything just went like machinery. I had a good gang of men. You take that little town of Trout Creek, you couldn't go up to Trout Creek and hire a poor man. They were all lumberjacks. And I had a good cook, a good clerk. The clerk had to keep track of all the men's time and he had a van with all kinds of clothes in, tobacco and everything. He had to keep track of that and take it off your time. Then he had to go to the cookery to order their supplies. Him and the cook would go around and see what they were short of, then when that came in he had to keep track of how much that all cost in order to get the meal costs.

There never was a company on earth that furnished like that Standard Chemical. The Royal York couldn't put up a meal like you'd find on those tables. That cook, Dave McLeish, could make a lemon pie and there wasn't a lemon within eighteen miles. That's the man that could make fifty pies an hour. As long as there's a lumberjack, Dave McLeish will never be forgotten. I don't care when you went into that kitchen, everything was spotless. He looked like a woman, just as fair as a lily. His shirt and apron would be snow-white. Not just white, snow-white. And when you sat down at the table you had something to eat. No woman could compete with his bread. And his raisin pie would be that thick!

I served a hot dinner in the bush. I got a sleigh made, cut down two maples about twelve feet long with a crook in them, put a box on it, two doors, and when you turned the doors back that made a table for your plates and cups and everything. I put a horse on it and had the cook go out to the bush with it. The hot meat, hot potatoes, hot gravy, everything was in that box. An old man had cedar blocks cut and set around, and every man had his place the same as at a table. He turned the seats all over so if it snowed everything would be dry. Turn the lids back, set the pots out, and you came around, scooped your meat, potatoes and gravy out, and kept on going. I had two kettles out there, and you'd throw your plate in that hot water. Now the dishes weren't washed, but all the dirt was off them to go back to the cookery. The first day was something terrible, you can imagine what kind of jam it was. So I waited till everybody was through eating and I said, "Boys, what do you think of the hot dinner in the bush?" "Oh," they said, "It's wonderful!" I said, "This'll be the last day if that's the way you're going to carry on — just going through one another, knocking plates out of one another's hands." You'd think they were starving to death, they were afraid there wouldn't be enough. From that day on it was just the same as it was in camp.

I only fired one man in my life. He'd go out on Saturday night, come back in with whisky and be no good for two or three days — and he'd want to get the same wages as the men who were doing his work. He'd bring in three of four bottles of whisky and hide them in the bush. He'd go to work in the morning and maybe an hour after, he'd walk from where he was working to where he'd hid the whisky and he'd get drunk while other men were doing his work.

I never let no old man go. I'd find him something to do, supposing he wasn't doing anything. I never would have remembered, but this old man came to the office and said, "I have to go out tomorrow." I said, "Why, is somebody sick?" "No." I said, "Dad, why do you want to go out? You came here looking for work and I gave you a job." "Yeah," he says, "but I can't stand the job. I'm shovelling snow off skidways and I can't stand that job, my back has given out." I said, "You stay in camp and I'll call on you." I gave him a shovel and an axe, just giping on the road, and he was there until spring.

ALBERT SCOTT (b. 1895)

I worked for my father for about fifteen years. His name was Thomas A. Scott. He was running camps all over the country. I worked for him in Killbear Park, skidding logs and drawing them, and peeling bark in the summer. He was taking out pine and hemlock. It was the same time as the Canada Wood, Lumber and Chemical Company took wood out; that would be 1913-16. That's the outfit that owned the chemical plant here [Parry Sound]. They were cutting just hardwood. They turned the wood into charcoal, and the juice out of it was wood alcohol. The idea of it was to get the alcohol, but they couldn't get the alcohol out of it without getting charcoal. There were a bunch of ovens and they loaded the wood on little cars. The track ran in under the door and they piled a cord on each one of these cars. After the wood was burnt it all went into charcoal, and the steam and fumes of it hit the walls and ceiling and ran down a crevice. They had a [distillery] down below — they used to build these plants on a side hill.

The wood had to be a year old so it would ferment. You can't make alcohol out of anything unless it's fermented. It had to be mostly maple. Cut and pile it for a year. They drew the wood down and piled it on the high rocks along the bay, then they'd take a scow and tie it to a tree or something, put a gang plank out and wheel it onto the scow with wheelbarrows. There were more than a hundred men over there. They cut by the cord, so much a cord for cutting, splitting, piling it and cutting a road to it. That chemical wood was cut fifty inches long and split so it would go through a six-inch ring. That was so that when they put it in the ovens the heat got through it.

When we were at Killbear it was W.J. Beatty owned that. The log stamp was IB, Isabella Beatty. Then there was a cartwheel. If they didn't have enough IBs they'd give you one of these cartwheel stamping hammers. I don't know whose it was; Beatty bought it from somebody, They bought out several limits for the hardwood that was on them, and when they bought the limits they bought the stamps that went along with them. Old Tom Woods looked after the bush end of the business for Beatty for forty or fifty years. He wore a King Edward whisker. He was what you'd call a walking boss for Beatty. They never ran any camps themselves, they got somebody to contract it. Then the contractor would never make any money because they just paid enough to entice you to start. They knew damn well by the time you got through you'd owe them money instead of them owing you. I never knew anybody that ever worked for Beatty that made any money. They made the money. All you'd have was the fun of the work.

WILLIS KENNEY (b. 1884)

Graves Bigwood was the biggest operator of a sawmill that was ever in this North Country. I travelled for a meat firm and sold half a carload at a time to the mill at Byng Inlet — three or four beef, six hogs, half a dozen cheeses, a quarter to a half a ton of lard, sausage and so on. Mr. Ritter, their walking boss, was an unsual man. He dressed very neatly, with high black boots polished right up, and he'd wear his pants down over them. He'd come to Billy Thorne to get his suits made. He had whiskers and he chewed tobacco. He'd come into the store and he'd have to go out a couple of times to spit. But he never discoloured his whiskers with tobacco like some people. He'd do the negotiating for a certain bunch of timber, and McCallum or some other contractor would get the logs out.

I was in McCallum's camp this side of Dunchurch, in behind the farms, one Sunday. The camp was very rough. The dining hall was just cheap lumber, rough boards, and just benches alongside the tables. The grub was pork — long clear bacon — and beans and syrup, either blackstrap or golden syrup. The bread was baked right there. The bunks were double bunks. I expect these men never had a bath from the time they went in till they came out. If there were men living in Dunchurch who had families, they would come out for a Sunday. But most of the men those days were floaters. They'd get into a camp, and I guess earned $15 a month and their board. A good man would get $20, a poor one $10. They'd come out in the spring with maybe $100.

In Parry Sound — my first business was in Parry Harbour, a grocery store — I saw the other end of it, when they came out from the camps. There was the Montgomery House and there was the Kipling. One of them had rooms especially for the purpose of rolling the men, getting their money. There would be men who would see that they got drunk, and others would take the rest of their money from their rooms while they were out. That actually happened. A good percentage of the fellows would come out in the spring, and unless they had families most of the money would go to the hotel.

EVERETT KIRTON (b. 1894)
When the Ontario Lumber Company bought out the Walkerton Lumber Company, I think that Albert McCallum was one of the men that was sent up. Whether he was a waif that they picked up or what, nobody seemed to know much about him. There was an old Ontario Lumber Company depot closer to Dollar's Dam, and I was told that was the headquarters when they were going to put the railroad through from Ottawa. A lot of the right-of-way was cut out and there were a couple of piles of steel in there somewhere, sunk down into the ground out of sight. They used that for the railroad headquarters, and the OLC used it for a while before they cleaned up the new depot.

That was where McCallum carried the mail. Once a week, every Monday, he'd make the hundred miles to Dunchurch and back in a day and a half. That was fifteen or sixteen miles north of Loring, and he'd be at Loring between eight and nine in the morning. It seldom varied over a few minutes. My mother used to look out to see him go by, and she didn't have to look very long. He went from there down over the old Pickerel Hills Road, below Arnstein, through to Glenila for dinner. Then after dinner, from there to Dunchurch, fifteen miles or so, get the mail and come back to Glenila. (Glenila was kind of a floating post office. Different people where the Pickerel Hills Road joined onto the Great North Road had it. It was Glenila Post Office whoever kept it.) Next morning McCallum would leave Glenila, arriving at the depot at noon. I've heard so many people talk about that. Jim Sinclair came out from the Depot Camp to Loring and he thought he'd come with McCallum. Jim Sinclair was a fairly young man, but he wasn't used to that speed. Mac had this lope, and Jim Sinclair couldn't hardly keep in sight of him, let alone travel with him.

JIM CANNING (b. 1872)
I run a lumber camp for Graves Bigwood. Bigwood was a jolly man, hardy, stout and well proportioned. He could handle the hands pretty good too, because at Byng Inlet he used to have lots of people that liked to get the gloves on. He was a mighty fine fellow. I knew Bigwood better than the other fellow because he was always around the mill. Graves was a real churchman, a nice man too. He was supposed to be the man that had the most money [in the company]. One time he was left at Byng Inlet in the fall, and Ollie Simpson and I were there. There was some long, twenty-four-foot hemlock that had been driven all the way to Byng Inlet. Bigwood was away and Graves didn't know as much about what was wanted, and he had Ollie and I cutting them in two. We would pull these logs in and measure them and saw them in the water with a crosscut saw. Bigwood came back and came right up to the [sorting] jack and says, "Jimmy, don't you know better than that? We got them to fill a bill, not to cut them in two!" "Well," I says, "I knew them cost a lot of money, but it's a good man that does what he's told."

ARNOLD McDONALD (b. 1907)
Sam Ritter [Graves Bigwood's woods manager] was a very particular man. He wore a grey serge suit all the time and his shoes was always shining. Tom Buchanan used to make the shoes, long pull-on, and they had quite a thick sole projecting out at each side. He wore a big long whisker which would blow like silk in the wind. People said he never shaved. He chewed tobacco. A lot of people without a whisker used to have tobacco on their chin. He'd cut it in little squares the size he wanted to chew and it was all wrapped separately. He'd take one of those out and put it in his mouth. And there never was a stain. He had a driver, Nell, and his buggy, then in the winter a cutter. If Ritter went out for a drive in the buggy, when he came back it had to be washed and shined just like you'd shine your car. And that

Samuel Ritter, his wife and daughter at their home in Ahmic Harbour. — RENA SAAD

Samuel Ritter, woods foreman for a succession of companies logging in central Parry Sound District. — RENA SAAD

Samuel Ritter, woods foreman based at Ahmic Harbour, with his team of driving horses. — RENA SAAD

mare had to be groomed just so. He was a very tidy man. The buggy had a top and a water-proof spread to fit over the dashboard, and to come back over you if the rain beat in. When he was walking from his place up to the office he had a pair of gloves, and he'd have one hand behind his back with a glove in it, and the other hand without the glove on his wrist. It was just his way of going.

JIM LUDGATE (b. circa 1900)

When my father was manager for the Schroeder Mills and Timber Company at Pakesley, he and Herb Thompson had a separate operation too. They had bought a part of the Schroeder holdings that lay down around the Key, a small operation, about 30,000 [board feet] a day or so. They started with the McKellar mill, that's going back sixty years or so, then they moved that mill to Lorimer Lake. From there they moved it up to what became known as Ludgate on the CNR, at the head of Key Inlet. They started, I think, in 1911, with their headquarters at what was known as Salines on the CNR. They had one camp close to Island Lake, and the logs came down the Magnetawan. Then they had one on or close to Noganosh Lake, and I think two other camps in Brown and Wilson. They were about four years on that limit. They moved the mill to Naughton west of Sudbury and operated for four years, I don't think very successfully. Then they had a fire and that wasn't very success-ful either. It burned the lumber yard and not the mill. They were done with the mill; it might have been a chance to make a little out of it.

It was 1917, I think, when Schroeder's started at that operation. They intended to be loggers only, but a mill had been built at Lost Channel by Lauder, Speers and Howland, and they hauled their lumber by sleigh out to Mowat on the CNR. When Schroeder's came into Pakesley, Lauder, Speers and Howland got after them to go in cahoots on a railway. One of them was supposed to be a railroad-builder. So they did. By the time they got the railroad built Lauder, Speers were broke. And Schroeder's, without a great deal of interest in doing anything except logging, were pretty well forced to take the whole thing over. They operated for nine or ten years. I think the last year was 1926. It was the end of an era in log-ging. Mechanization was just coming in. The Schroeder people — they were American, from Milwaukee — were all for Linn tractors. They were caterpillar behind and skis in front. My dad said, "The day you arrive here with a Linn tractor I'll leave." My dad was very fond of horses. The bulk of the Schroeder holdings lay up the Pickerel. There were some terrific stands of pine. They cut eight winters within walking distance of one camp of 125 men.

Schroeder's had mills in the States. A lot of it was beer money, this Schlitz beer you see advertised. Uhleins were the Schlitz people. They had money to get rid of somewhere, I suppose, and it was through the Schroeder outfit. Any of the old-timers around will tell you what a terrific temperance man my father was — and him working for beer money. There were none of the Americans over here, it was practically a hundred percent Canadian oper-ation. They had the money, but they seemed satisfied to leave Dad to run the show. The managing director of the firm would come over once or twice a year from the States, but other than that it was run entirely by my father. Most of the lumber went to the States. I started to work about 1919, briefly, then I worked a couple of summers while I was at uni-versity. I took civil engineering. I started working for my dad at the headquarters in 1923 and stayed there till it ended, in 1927 I guess. Then my father went in with J.B. Tudhope of Orillia, upriver from Maple Island. They had a sawmill on Gooseneck Lake. It ended with 1929. They crashed. You couldn't sell the best pine you ever saw for what it cost to put it on skids. And it cost quite a bit. I worked there for a couple of years, but the crash came and I had to get out.

CLIFF BENNETT (b. 1909)

My dad remarked to me several times that the roughest, toughest river-driver or lumberjack never called Mr. Ritter [woods manager for Graves Bigwood & Co.] anything but Mr. Ritter. Nobody ever used Sam or anything like that, they had too much respect for him. He was respected; he was Mr. Ritter.

GORDON WHITMELL (b. 1899)

I put in four winters for the Schroeder company up at Pakesley. Everett Farley was foreman. The first winter I was at Camp Three at Smith Bay, and I worked three winters at Camp Seven. We went in about four miles from Pakesley and turned left, and both Camp Four and Camp Seven drew down the same log road to where we dumped at Kidd's Landing on the Pickerel River. I liked Everett Farley, he was a quiet man and good. When I went up there in the fall of 1921 wages were pretty low -- $22 to $26 a month. It was that summer Dad built the barn and I told him I was going up to the camp when I got the ploughing done. "Aw," he said, "You've worked hard all summer and for all you'll get you might as well stay home." I said, "I'm going up there for a rest." I went up to Pakesley from Parry Sound on the train, got in there at ten o'clock, and went in and found a bed in the bunkhouse. Next morning after breakfast I went into the office and asked if there was any chance of getting a job, and he says, "You can go in to Camp Four to Orv Gray. Do you know him?" I said, "Yeah, I know Orv Gray all right, but I'd like to go to Camp Seven. I worked for Farley for three winters and I'd like to go back again." He said, "There's no use you going in there, he has more men than he knows what to do with. There's men sleeping on the floor." They gave me a ticket to go in on the train to the storehouse, and after a while I thought, hell, I'm not going to wait for the train, I'll walk in. I was pretty sure Farley would give me a job. It was seven miles from Pakesley in to Camp Seven, and I walked in, got there about eleven o'clock and went into the camp. Farley was there. I said, "Any chance of getting a job here?" Farley said, "You damn right there's a job for you!" The scaler's helper was away and the first two or three days I helped the scaler. Then one morning I was coming out from breakfast and Farley called me from the office. "Come here, I want to see you." So I went in and he said, "How would you like to take that bay team out this morning?" They were just a new team that came in and they didn't look as slick and fat as the company horses. I said, "Oh god, I don't know; I've always been used to a good team." I said, "There's another team -- that fellow who is going out this morning, why not I take them?" He said, "I know, but the other team, if I don't get somebody who can handle a pair of lines, they'll soon be no good to anybody. If you can take them and make a team out of them by Christmas, I'll see that you'll get a little more money." I said, "Well, OK, seeing it's you I'll take them." They were a well-broke team, but they'd never been in the bush. They'd go straddle of a tree or jump into a brushpile or anything. Anyway, I took them, and after about three days I wouldn't have traded teams with anybody. They just suited me, a pretty high-strung pair, and I liked them. So he gave me $35 a month at Christmas and I took my own team up after Christmas. I had just a front sleigh and a box on it. Some hay in the bottom, and the horse blankets. I went up between Christmas and New Years. It took me three days. I went from home [Dunchurch] up the Nipissing Road to Commanda the first day, then Loring the next day, and a day from there in to camp.

On the sleigh haul at McCallum's it was three o'clock when we were called in the morning, but up at Pakesley they got up at four for breakfast and the lead team would pull out about six. But that fall we'd hear Orv Gray's men at Camp Four chopping in the bush when we were going for our breakfast at Camp Seven. They had to come that mile, mile and a half, past our camp, and lots of times his teams would be going down with a load before our lead team would be pulled out. Both for the same company. And he didn't have any more logs skidded than Farley had. He was just foreman, getting paid by the month. He was a nice chap, I liked him as far as meeting him and talking to him. Orv Gray was all right. But they called him "Moonlight" Gray.

JACK CAMPBELL (b. 1887)

The Graves Bigwood had the darndest payroll you ever saw. There was no government spending money among the people. If you didn't have a job you could starve. But if you worked for the Graves Bigwood and something happened to you, they never took you off the books. They kept you right on. An old man that would get so he was able to do nothing, they kept him just the same. I don't say they paid him as much as that man that's working, but they kept him from starving. He was still kept by Graves Bigwood. Old "Major" Anderson couldn't do nothing. They sent him in to be choreboy. Well, he couldn't chore no more than — but [he had] a wonderful memory to tell a story. And he wouldn't tell any stories without there was lots of wood in the yard and water carried in for the cook. We fellows, we would come in, start gathering wood, carrying water. There's nothing else to do, so you might as well do that. Then the old fellow would start into one of these long stories. Maybe it would take a week to tell one story. Very old-time stories. Charlie McGhie was telling us they went in to scale, so he started telling them this story. There was always in it this fluffy little kitten. Charlie says, "That story lasted all the time we were scaling them logs, and all there was to it, at the last, was that this woman wiped herself with this here kitten." That finished the story.

ED PLETZER (b. 1887)

The last camp I worked in was at Dog Lake, for the Parry Sound Company. It was a big camp, about a hundred men there. Eli Mairs was the foreman. Another fellow was foreman for three years before that. The boss came out to the bush one day and this foreman was chopping and measuring the logs, and I guess he'd cut a little piece off the end of his measuring pole. The boss measured his logs and they were all two or three inches short. He sacked the man and Eli Mairs got the job.

Eli wanted me to go in for him that winter. I was rolling, then after Christmas I was loading, sending up. Rolling seems to come natural to a man if he's leaned that way. I had a good man with me for sending up, Charlie Schell. We were good chums and we used to try to get loading together. He was right-handed and I was left. We had a good top-loader and we could put up loads faster than any other gang in the bush. We loaded one winter back at North Lake, for Gould and Thompson. They had all short haul, onto Jeffery Lake and North Lake. We could see the team on the lake with a load when we were starting to load another load. And Charlie Schell and I kept them going all the time. Alex Farr was our foreman and he was about the best top-loader I ever saw. He had two loading gangs with two teams drawing out in each gang. The other gang hadn't one load out yet, and he came to their skidway and says, "What's wrong?" This fellow said, "There's nothing wrong." Alex said, "There's nothing right. There's two teams gone in the other gang and you haven't got one load out yet." Alex says, "Come down out of that." And he got up on the load in his place and got Charlie Schell and Ab Johnson for sending up. They says to Alex, "How many logs do you want?" And he says, "All you can send!" They put the chain around five or six at a time.

Charlie Schell and I used to put the bottom tier on in one draw, fill the bunks. Something you've got to learn is to put your decking line in the right place to keep the log from sliding endwise -- that's a "gun." I got taken right off a load that way once. I was top-loading and they sent a log up, and when it got to the top it swung around. I caught the end of it with my hook, just caught the front of it, and it took me and that big log right down onto the road. I still had ahold of the log when I landed. If I hadn't caught it it would have broken my legs. It was the teamster's fault; he was letting his horses go too fast. He had a big team and he was drawing up a grade, and they went up like a shot.

One time I was loading and the boss was there. He was going to help send up a big log, a peeled hemlock twenty foot long and twenty inches in diameter. He put the line on it three times, but it would get up halfway and the chain would slide -- there was a lot of ice on the log. Charlie put the line down again, and I took it this time and put it around the log twice. I handed it up and I says, "Get your dog in and get the line good and tight, tight from your dog down to the log." It went up the first time. I'd put the line around twice, you see, and that kept it from sliding. The walker for Peter's, the walking boss, he was standing there watching us load. He said, "I never seen than done before." Oh, that Charlie Schell and I would load against anybody. We didn't make very many mistakes.

A boom of logs at Good Cheer Island, Georgian Bay. VH carved in the boom shows it belongs to the Victoria Harbour Lumber Company. — JOHN ANDRAS

Part Two:
GOING ON THE DRIVE

POWER AND BALANCE:
RIVER-DRIVERS

JIM McARTHUR (b. 1886)

In 1904 or 1905 this young Parsons and I made arrangements to come up from Toronto to Parry Sound to go hunting with George Sirr. We came up the Grand Trunk to Scotia Junction, then in on the Canada Atlantic through Sprucedale.

We went up above Mountain Chute to a rough board cabin that was built for the river-drivers coming down -- they used to have a lot of trouble there. I was sitting on a deer run where the deer crossed just above the rapids this day, and they'd been talking about Charlie Reekie, who had been around the previous day. I was sitting there and all of a sudden I heard a sound. I turned around and here's Charlie Reekie. He had a big black hat on and so many whiskers all you could see were his two eyes and the tip of his nose, but I knew him as soon as I saw him from being told about him. He wasn't tall but he was broad. He had moccasins on, I guess he'd made them himself. He had his gun. He came down and started to talk to me. He said, "You're with George Sirr, are you?' I said, "Yeah." "Oh, George is a fine fellow." George was a great booze artist, and I understood that if Reekie was like most of the rest up there he was a booze artist too. So I said, "Yeah." He says, "Tell the boys I called." I knew damn well he wanted me to go to the cabin and give him a drink. We talked for a while and he told me about river-driving up the Seguin, and away he goes across the river. He didn't wade, he stepped on the rocks -- and some of those damn rocks were twelve feet apart! He just seemed to spring and he was across. Never wet a foot, I don't think. Four or five jumps and he landed on the other side, and away he went. I thought I wouldn't like to have that man after me.

Reekie was strong, a powerful man. On one of the last big drives on the Seguin they had a warp, a thing with a winch on it they were going to take down the river. A lot of the drive was over the Mountain and down toward Burnsides. These fellows were on this crib and

111

arguing, three or four of them. There were two anchors on this raft and they were arguing, "I bet you can't move that anchor." Charlie came down and asked, "What are you arguing about?" "Well, we were just betting if one of these fellows could move one of those anchors." They were heavy and long, different than a naval anchor, but they gripped the ground, gripped the rocks when they went down. They weighed 600 pounds. Charlie says, "How about getting a bet?" They says, "Sure, it's a bottle of whisky when we get to Parry Sound." He says, "Make it a gallon of whisky." They knew he was strong all right, but they laughed at him. He got that anchor up on his knee and he says, "There you are!" Then he took it over and threw it in the Seguin River, and it's there yet. That's true! A fellow that was on that drive said it was true. He said that man never knew his own strength.

You've heard about the time the guy tried to pass Reekie with a load of hay. He was coming from his place with a load of hay for Parry Sound and the road was only one wagon wide. He got about a mile towards Parry Sound and another person was coming along in an empty wagon and wouldn't give way. Charlie said, "You're empty, turn into the ditch. We'll get you back out." "I've got the right-of-way," the other fellow said. He evidently didn't know Charlie Reekie. Charlie got off his load of hay, walked up, grabbed the guy by the collar, pulled him off and shook him a bit. "Now will you move your wagon?" The fellow said, "Yes, I'll move it, but you'll hear about this. You don't know who you're talking to." Charlie said, "I don't give a damn who I'm talking to, I've got the right-of-way; I've got the load and you haven't." But sure enough he lodged a complaint. There was no policeman in Parry Sound, or if there was you only saw him occasionally. They had to send this Greer up from a bigger town nearer the city. He landed in Parry Sound, got a horse and buggy off someone, and away he went. Well, Charlie was plowing down from his house quite a ways, putting in a crop. This guy had been asking along, and he came to the place where Charlie was plowing, and he says, "Hello, can you tell me where a person named Charlie Reekie lives?" Charlie picks up the plow and points at his house, "Right over there!" Hah! And Greer turns around and goes back to Parry Sound. Well, that's the truth; that's no fable.

WALTER SCOTT (b. 1893)
I guess Charlie Reekie was the most powerful man around this country. Charlie's legs were short and stout; down at the boot tops they were bigger than any place on my leg. I skidded for him one fall when he was getting to be a pretty old man, in his seventies, and he would pick up some of those hemlock logs and throw them up like they were handspikes. No trouble to him at all. There was a gang of men putting on a load of logs and one great big log went off the bunk, so the three rollers got a hold of this log with their canthooks and tried to lift it up. Charlie says, "Just step out of the way, fellows." He picked up the log himself and threw it on the bunks.

JACK McAULIFFE (b. 1901)
Most of the drivers on that drive had come from Montebello, Quebec, French-Canadians. The older ones were very experienced and very skilful. One fellow, Legault, for what you call the sweep, had this particular big pine log that he used. As the logs moved down, some got caught along shore, and the last was the sweep, where they went along with these pointed boats with four oarsmen and two paddlers in each. But he used this log, went along with his pike pole and tied it up at night. That was his boat. He took it down as far as I went, anyway.

DON MACFIE (b. 1921)
Dave Wye could never learn to drive a car. You see, he spent years as wheelsman on an alligator on the log drive, and on them, when you turn the wheel left, the boat turns right.

WILLIAM JOHN MOORE (b. circa 1895)
One time when we were driving square timber down the Magetawan we camped at a falls. Near morning "John Dollar" [James Madigan] asked me if I had noticed anything odd. He

said he had dreamed about some man accusing him of sleeping on his grave. Then when daylight came we found we had slept on top of an old grave.

I used to ride log rafts across to Michigan as a watchman at a dollar a day. After the government stopped them taking logs out of the county, I helped steal a sawmill in Michigan — though I didn't know it at the time. The Peter Lumber Company sent a tug across to pick up this dismantled mill, which we heard later didn't belong to Peter's.

HECTOR WYE (b. 1905)

Uncle Bill [Wm. John Moore] used to be log man on one of those tugs. They had to give the boom over to an American boat where the line was supposed to be. He said, "The last trip of logs we took, we brought a sawmill back." I guess that was when Ontario cut off towing logs; they had to be sawed here in Ontario.

One time the captain, the guy that owned the boat, had his daughter on with them. They just had a couple of lanterns hanging on the boom, that's all they could see of the logs coming behind at night. It couldn't have been too far out from Parry Sound. He was standing there on the back having his evening smoke before he went to bed, and he said the back of the boat seemed to raise way up, then it settled down again, and the lanterns were still coming and everything looked to be OK. But in the morning they found they'd lost the logs. They still had the storm booms, but that's all they had. They'd opened somewhere and let the logs go. I think the logs were still in their own booms, but had got out of the storm booms. The owner was in a heck of a way because the logs had gone into the shore. He went to Parry Sound and got a bunch of guys to help. They weren't drivers or anything, didn't know very much. They spent two or three days and still weren't much further ahead getting these logs gathered up. Uncle Bill thought he knew how they should be gathered up. The owner wouldn't listen to him, but the daughter seemed to think he could do it — I guess she had given up hope of of doing it any other way. Bill said, "I think we could get the stuff gathered up if we had some way of keeping your dad out of the road." She said, "I'll fix him. I'll lock him in the cabin and you can go ahead." So she locked him in the cabin, and in less than a day they had the logs back in the storm booms, and away. Uncle Bill was pretty religious, but he told that story so many times that he thought it was true.

July 12, 1888: Blowing fresh all day. The Superior came in; she lost her tow of 12,000 logs in the Bay and came very near swamping in the gale. — D.F. Macdonald

ARNOLD McDONALD (b. 1907)

You had lots of spare time, and Mr. Harrison said to me one time, "Did you ever think about walking that tow rope?" It was about 250 feet long, to keep too much current from hitting the logs. From a steam tug there's quite a shove of water, and the tow post was quite high. I said, "No, I never." He said, "I'm sure you could walk it." He had been on there for Graves and Bigwood for years and years, and he said, "There was only one man in all the years I've been on here, Ed Bottrell, he walked it." So this day we were through Deadman Narrows. I had the punt back there, and the water was warm then and I could swim, so I said, "I'm going to try that rope." I took my caulked boots off, left them in the punt and went in my sock feet, holding my pike pole crosswise for balance. For the first few feet coming from the logs my feet were in water. I made her. Jack O'Halloran was running the engine. I told Jack and he said, "You never did!" I said, "Well, I'm up here and I'm dry and the punt's down there." And he went out of there and up into the wheelhouse and told Mr. Harrison, "Arnold walked the rope!" Harrison said, "Well, I told you Arnold could walk it." Jack said, "How are you going to get back?" I said, "I'll walk it." He was down in there monkeying around, so I thought, I'm going to take off. You're quite high up, you would be ten feet or more up where you started from, and when I was halfway down he looked out and saw me and he shut the steam off. The rope went that way, then I went this way! And Mr. Harrison said, "That was a dirty trick, Jack. He would have made it." Jack said, "That's why I shut the steam off, I knew he was going to make it!" He was a comical bugger, Jack O'Halloran.

ROY WAINWRIGHT (b. 1918)

I drove the Magnetawan to Byng Inlet two springs with Dougald Campbell. There were thirty-five or forty men on these drives. You eat four times a day, because you take advantage of your wind and daylight, and the days are long. Dougald Campbell was very superstitious. He wouldn't tie a boom of logs to a poplar tree. He claimed that Christ was crucified on a poplar tree. But I think the reason for that was that there are no roots on a poplar. They're just like a big milkweed, no root to them. If you've got a big boom of logs and want to tie it to a tree so it won't get away in the night, you don't tie it to a poplar, you tie it to a pine or yellow birch, something like that.

Another superstition he had, he wouldn't hire a McDonald. Maybe you've heard the story, supposed to have happened in Scotland. I remember some of it from when I went to school. It was in Scotland, where they had clans, and the McDonalds and Campbells were at odds with each other, but they formed a truce. Then the Clan of Campbell came to the Clan of McDonald to stay overnight in the hills. I remember the chorus of the thing: "And they murdered the house of McDonald...." I think this is the reason Doog wouldn't hire a McDonald.

We were down at the Fourteen Rapids, down near Byng Inlet, and this day when first lunch came along, Campbell said, "Come on, young fellow, get your axe and come up on the rock. We'll set up a dinner place for the boys, get some dry pine and boil two or three pails of tea." He'd always take one of us young fellows to do that, because the older men were more good to him on the water. And he was sitting there telling me a long-winded story. He was a great horseman, and he believed there was a superhorse that could fly all over the other horses. He'd seen it going. He was right in the middle of telling me this ghost story, looking into the fire, and all of sudden the fire went all over the place and the tea pails upset. Well sir, the old fellow started running and I passed him. To make a long story short, along the river there are those mud turtles with red spots on their shells. It was hot weather and some had crawled in between the moss and the rock. We had built our fire over top of them, and when they got hot they took off and took the fire with them. It scared Mr. Campbell quite bad.

Chris Carlton used to go to town every weekend and get drunk, and on Monday morning he was just like a bear, cranky, miserable. There was a bunch of logs on this river drive that fire had gone through and killed the bark on them. Some had floated out of the bark and left the shape of the log in the bark. Every morning they would bring the tug to the side of the boom and let the men off. This morning a bunch of us young fellows — Albert Bottrell, Arnold McDonald and I — took a bunch of these logs that had floated out of their hides and put the bark where we knew Chris would run across to get to the sorting jack. He ran out and there was nothing but bark there, and down he went, flailing and swearing. He was a terrible man to swear. The boss conveniently didn't see it.

JIM McINTOSH (b. 1896)

I remember a fight at the Kipling Hotel one night between Dutchy Herring, the Bully of the North, and Black-eyed Shackleton from Sault Ste. Marie. Herring was born in Sebright. He was chucker-out in the Atherly Hotel the night the army was there — the night they smashed every window in the hotel and he was the only man there wasn't a mark on. He wore a pair of caulked boots. This Black-eyed Shackleton, that night he said, "I'm the best man in Parry Sound." Herring said, "You are when I'm in Sebright, but when I'm in Parry Sound I'm the best man." Then out the door and at her they went. There was a white board fence, little narrow boards, right down one side, across the end and up the other side. I don't think there was a board on the fence there wasn't blood on. When it stopped Herring was laying on his back and Shackleton was laying on his side about four feet away. Shackleton rolled over and said, "Herring, you're a god-darn good man." And Herring said, "Don't talk so foolish. I got six brothers in Sebright and I'm the poorest one of the bunch," and jumped up and at her again. Herring was a friend of mine. He said, "I hit him hard enough and often enough to kill a two-year-old bull." That Kipling, at one time that was a terrible

This immense bank dump is believed to be on the lower Magnetawan River. — JAMES T. EMERY

Stranded log drive on the lower Magnetawan River. — JAMES T. EMERY

place. A lot of good men went in and out of that place, old river-drivers and old lumberjacks.

MARSHALL DOBSON (b. 1891)

Jack's Dam on Steidler Creek was right close to Number Two Camp, so at night they would get some horse manure from the barn and stuff it between the logs so it wouldn't leak. There was always a thick layer of soaked horse manure on the water above the dam. This day, George Bayne and I were tailing logs through, one on each side of the gate, and a big log stuck against the dam on George's side. George kept poking at it, and finally got mad and took a swipe at it with his peavey. He missed and fell in. He came out all dirty and headed for the camp. McCallum came along, saw what was what, and he tried to get the log out, with the same result. He took out for the camp behind Bayne, just as wet and dirty.

Ben French, he was good on the logs. There was a big fellow, Buchanan, he was extra good, only it took a big log to hold him. He weighed over 200, an awful strong man. They would often ride a log down through a dam. You couldn't knock Art Macfie off a log either. "Scobie," that was the name they gave Art. It was McCallum who started it. Art started working for him quite young, and on the drive he wore this great big pair of hand-me-down boots. When they got wet they curled part way around any log he was standing on. McCallum said he hung onto the log like a Muscovy duck, and that's where the name Scobie came from.

JIM CANNING (b. 1872)

We were running the [Mountain] Rapids and Aaron Teneyck says, "Jim, you get in the punt and we'll run the cookery over." I says, "Gol, you're going to drown something." "Oh no, maybe not." This boy was in the punt, and the cook and Aaron Teneyck and me. There was a barrel pretty near full of sugar, and one thing and another. All the stuff we had for eating was in this one big punt. It went in a little crosswise. There were lids off the stove, and this young lad stuck his head down there, stuck his head in the lid. I can see him yet! The worst darn place he could be if it was going to turn over, with his head in the lid.

That Seguin drive, boy, that's a long time ago! Just count back — I must have been sixteen when I was down there. Sixteen off of ninety-two, what would that be? Seventy-six years ago.

September 20, 1893: The Georgian Bay L. Co. lost a tow of 5,000 logs on their way down from French River last Saturday. —D.F. Macdonald

River driver at the foot of Mountain Rapids, 1885.

BIRLING

JIM McINTOSH (b. 1896)

My brother-in-law could take a log and lay right down on his back flat, with a pike pole across his chest balancing himself, and he could get up on that log — a round spruce log sixteen feet long, a log that will [just] carry you standing on it. I could lay down and ride all day, but I never could get onto my feet, and I tried her and tried her. He knew lots of tricks. Take a cedar tie that'll carry you in the centre, take your pike pole and stick the gaff in the end and keep going back. The tie will keep going down, down, down. Keep pulling on here, turn that right on end, and then step around and he'd be on the other side when she popped up. I've seen him do that in his bare feet. Bare feet will stick to cedar bark good.

ARNOLD McDONALD (b. 1907)

I was light and I could really go on logs, take one and paddle it anyplace. And I used to run some through the rapids. I went through Lovesick to show that I could do it. Somebody might say you couldn't. I didn't always stay on, but quite often I did. You want a log that floats about six inches out of the water. You had your pike pole to balance with.

We used to have birling contests among ourselves. The best fellow I ever saw at birling logs was Alex Beausoleil from Penetang. I birled with him, but he was just teaching me. I never would have been able to stay with him. He could take a log and it was just like white water falling, it was going around so fast. I birled with a lot of people, but the only one that could wet me was my brother. He could eventually get me. You would try to stop it quick, jump on the other side and stop it. You took a log that was just big enough to hold you up, because a big log is too hard to handle, and there's too much of it out of water. You'd stand about three feet from the centre and both start it, have your pike pole up in the air out in front of you to balance yourself and start it going back with the toes of your boots. You were barely just jumping up and down, one foot then the other. But he could get me in the end. If he got it stopped too sudden on me, I went in. This one time I knew he was about to get me and I just stepped up a little towards him and grabbed him and took him in with me. He didn't like that.

There was a knack in running logs. What you do is watch your legs. You step on every log you come to. Supposing you run into a small one, you're just on it and off it again. You're taking short steps. I've seen fellows that figured they were pretty good, and they'd take a long step, and before they got their other leg trailed up there, that one was down, and it wasn't long until they were down on hands and knees. Step on every log you come to. You might come to a place where you have to speed her right up to get off the other end before it goes down.

ALBERT SCOTT (b. 1895)

I never was a river-driver. I couldn't swim and I did well to stand up on the township without standing on a log. But those Campbells — Doog Campbell's family at Waubamik — all those fellows were river-drivers. Jack and Dan and Mac and Charlie and Harry, the whole outfit of them were good on logs. They lived right there by Land's Creek, and every once in a while they used to draw up a pine log, put it in, get down there with pike poles and practice birling this log. It's no damn wonder they were good at it, because that's what they did for a pastime. They could do all the tricks that are available to do on a log. Jack was the best. There were three Jacks; there was a hell of a tribe of Campbells around Waubamik. Doog's Jack, or Gentleman Jack, is the one I'm talking about. He blacksmithed in McKellar and I blacksmithed for him in the summer of 1916 after he bought Tom Canning out.

Another man that was awful good on logs was Frank Green. They lived up above McKellar. He could take a fence post that was eight feet log and a foot at the top end, and get out on it with a pair of caulk boots, lay down on his back and sun himself on the water. He'd get in a bay where there'd be a log that was good to birl, a nice round one that would go to beat hell. He'd get up on this log and stick his pike pole down on the bottom, lean it up

Believed to be the Three Snye Rapids in the Magnetawan River, about 1900. — JAMES T. EMERY

A 'centre jam' on the lower Magnetawan River. — JAMES T. EMERY

against his shoulder, and stand there and birl that log until he'd have the bark chewed down to the wood with them caulks. You'd think it was somebody cranking a grindstone too fast, the water spinning off it as high as that doorknob. The bigger [the logs] are, the more they throw water. A greenhorn, if he didn't know better, he'd think if he got on a great big log it wouldn't turn with him. But the bigger they are the faster they roll. You see, the circumference is so great that if you get it turning at the same rate as a little one, you have to run to keep on top of it. And by god, a log that's lying in the water is awful easily turned, especially if it's nice and straight.

The bigger you are the faster you can birl. It takes weight in your legs to get them going. But when they do get going, they're darn hard to stop [unless] you know how to stop them. All you do is jump up in the air and come down on the side of the log with one foot. Jump in the air and turn facing endwise of the log, tramp the one foot in the outside of it, and that's like a crank on the outside. If you try to birl it the other way, it takes a long time to get the weight of the log stopped. Just jump on it and drive this caulked boot into the side that's coming up, and that hinders the speed of the log until you get it stopped.

DAN CAMPBELL (b. circa 1880)
My Uncle Charlie Campbell was reckoned a good man at birling logs; then there was Neil Campbell, he was good too. My dad [Dougald Campbell] was quite a man to drive bad creeks. When you're driving there's what you call a back tow. The water will run down that side and back up the other, and the logs will get in there. He'd catch boom timber and fix that, spike them down so the logs would run straight on down the creek. That saved a lot of work for the men; you wouldn't have to sweep the logs. In a head wind we'd work on these side piers, straighten the creek. [Where log jams might occur] he'd have a switch boom, so if the logs came too fast, why you'd just throw the rope over to the other fellow and he'd draw this boom across to hold back till the creek would get clear. If it was real wide they had two booms on.

GUY SMITH (b. 1885)
I never bothered fooling around, I never took any chances, but lots of fellows rode logs through rapids. When they were sweeping a rapids they'd keep a fair-sized log, and when there were no more left they'd roll it off and away they'd go down the river. There's no steering to it, just jump on with your peavey and away you go. They'd go down the rough water and when they got down where there were more logs, more work to be done, they'd jump off in the river and walk ashore. If it was white water they were all right, because they had their peaveys with them. If you're wading in fast water and haven't got your peavey, you'd be swept off your feet. Supposing the water's up to your chest, you hold your peavey upstream, down to the bottom. Your peavey cuts the water so you don't get the full pressure of it. Hold it right to the bottom, that takes care of your balance. And when you move your peavey, move it fast.

Running boats in a rapids, you go down and have a look before you run it. You can tell by the water where the best place is to run. If you don't get the smoothest place, the boat's liable to fill on you. Maybe over here the water's jumping four or five feet high, but if you get a smooth spot, take that and you'll get through without any trouble. On the Spanish River I had two fellows running the boat, and they were the best I ever seen at it. When I wanted to portage they said, "Oh, we want to run her." They'd go and have a look at it, and have no trouble at all. They'd run right tight to the shore; the water's not quite so strong there, or rough.

The Campbell boys were the best I ever saw for doing tricks on logs. There were a lot of Campbells and they were all good drivers. They used to come here to Parry Sound on the first of July, the big day, and show them how to birl — birl a log backwards and forwards. Jack Campbell was the best. The old man himself was good; a great old bushman, Doog Campbell. You know, Parry Sound used to be full of lumberjacks and drivers, and real good ones. You can't find one now, not one. There's not one here now that was in the bush when I was. There isn't one you can talk to.

A bank dump on the Pickerel River. — JIM LUDGATE

The Pickerel River filled shore-to-shore with sawlogs. — BILL LITTLE

A FAIR WIND

DAN CAMPBELL (b. circa 1880)

We didn't mind working on the drive. A fair wind, you worked hard; a head wind, you rested. Sometimes you might be a week you wouldn't do a tap. You'd get up every morning, but if the wind was bad you wouldn't waste the water. Hold back. Soon as it turned fair, you'd work long hours. They had to watch the wind and the water. The way they used to handle the water, they'd go [to a dam upstream] at three o'clock in the morning and put the water on, and when it came daylight everybody would be at work. It might take three or four men to open a dam. A big spool on top lifted the gate. All these big ravines had a dam on them that held water back. When the dams used to be full you could take a shovel and dig anywhere and get water. You can't do that now, the ground has got dried out since that.

The last time we drove down to Parry Sound for the Parry Sound Lumber Company, we were in Mill Lake for two weeks. It was on in the summer and they were storing the water, saving it for the [electric] lights. So on a Sunday there was a real great wind blowing, and my father said to Reece Hall, the head man of Parry Sound at that time, "If you'll let me drive today, why we'll be out of here." So they opened up the Mill Lake Dam that morning, and of course it was after dark when we got done at night, but the drive was right out into the Sound, into booms they had hanging there past that big high railway bridge. The gates [in the generating station dam] were quite wide and there would be seven or eight, maybe ten or fifteen logs going through at one time.

ERNIE CARLTON (b. 1891)

On a certain day in the spring, McCallum used to figure that whatever wind was blowing that day, there'd be an awful lot of it that year. He'd get up early in the morning, and if the wind was coming from the west — of course the rivers mostly ran west — it would give him an idea, figure it out some way, how long it would take to get the drive down. If there happened to be an east wind, the drive would go down a lot faster. Sometimes they didn't even get the drive right through.

WILLIAM McKEOWN (b. 1885)

Dick Robinson jobbed for the Parry Sound Lumber Company. At the time the sun was supposed to be crossing the line [the vernal equinox] he would sit here in McKellar with a knife in his hand, and a stick, whittling it. He'd whittle away at that stick and wait to see how the wind would blow. That was supposed to be the prevailing wind for the summer. Which it is. Well, he'd whittle, and if it was blowing from the north he'd grab his team and away to Parry Sound, take a contract for the drive. And sure enough, it would blow him right down to Parry Sound. Drive her awful fast; every other day would be a fair wind. He always made money. There was a gang from around McKellar that always followed him, Jake Fletcher and some of the Moores, and Charlie Harris. And Quackenbushes, great big fellows who came from Pembroke. He was a good man to work for, finest man you ever struck. Very best of board, you could be sure of that. He did some logging for himself at Oliver Lake. He got the boys to all take up homesteads in there, and he got quite a few hundred acres. He had quite a cut of hemlock, peeled the bark and drove the logs. Peeled hemlock is great stuff to drive. Your breath will make it go and your breath will make it stop.

Canal Rapids on the Magnetawan River, with its sheer rock walls.

BOYS ON THE DRIVE

DUNCAN CAMPBELL (b. 1893)

I first went on the drive when I was twelve years old, working for my father [Donald Campbell] on the Shawanaga River. All the young whifferts they could gather up were on there. I just jumped around like a rooster. I'd get out and feed the boom into the dam, poke the logs into the dam, and run the logs. When the logs jam up you got to run across them to get them loose, make them move up and down.

The Shawanaga was kind of dead water, only just a play toy, a lot different from what the Magnetawan was. One Shawanaga drive took six months. You'd be afraid to pick your pole up for fear the gaff would fall out of it, because it was dried right out from laying around. Head wind all the time. The Shawanaga is crooked. You don't go far till you find a crook and you can't go anyplace. When we were windbound we did everything we could think of: run, hop-step-and-jump, wrestle.

I got my caulked boots in Parry Sound. Butler made them, Sam Butler. You went to him and ordered what you wanted. Mine were three-quarter length, about six inches high. They used to laugh at me. They'd say, "Come here, I got something to show you." They'd say to me, "Stick out your foot." Just a little wee thing. My boots were two's. They got a great kick out of those boots.

My brother Donald had Dad's team on the Magnetawan drive. Lots of times he'd have to make a circle. There were different places when he left that we never saw him for a couple of days. But he'd get back to the river. When I was fourteen they put me down over the rocks at the Canal Rapids on the Magnetawan. You had to go down and break the logs off the centre rock. Every year there was somebody who had to go down. An Indian went down with me. The reason they picked us, we were light and easy to pull back up. I only weighed 120 pounds and the Indian was a little heavier. They had the warping line, a big rope, with loops made for you. They strapped it under your arms. My dad told me, "Be sure and fix that rope good around that Indian. Get it wrapped around him so that we can be sure of him." They'd just give you an inch at a time. There were fellows over on the other side of the channel yelling at you, telling you what to do. I knew what I had to do anyway because I had watched other fellows go down. You touch the rock only just to save yourself. The Indian didn't put his feet against the rock, and he was bouncing all over and got scratched up. All you had to do was dig the pile out. Once you loosened up one or two the water would get ahold of them. We might have been there two or three hours. It's an awful place for fast water, but it was all right for us. There was a big hole in the back as big as this room, where you could get away from the water. They'd leave peaveys down in that hole till the end of the drive.

JIM McINTOSH (b. 1896)

I started on the drive when I was seventeen. I've waded up to the neck from daylight to dark and snow on both banks. You'd wade so long your legs would get so numb you'd have to go to shore and run up and down as hard as you could go. Then hop right back into her again. We didn't have a fire at night. You went to the tents and jumped into bed with your wet clothes and steamed yourself dry. There was no camp with a big box stove, just tents along the river bank. What good would a fire be with a hundred men and the wind howling cold and maybe snowing? The best thing was go right to bed after you ate. Where will you find men today will do her?

You couldn't discourage a boy in those days. Boys those days — sixteen, seventeen, fifteen, fourteen — if you tried to make people believe the things they did, they'd say you were crazy. They were the ones a good foreman took pride in. He knew they were coming men. Like old Jack Snell, old Orrin Tolman, old Nels Tolman and what they tried to do, tried to make men out of them. And they did. They didn't have to show them very much, their fathers had showed them before they left to go to the camp. I knew as much when I went to the camp as pretty near any man there. Why? I'd been in the bush with my dad since I was able to run around.

ANDY AINSLIE (b. 1889)
I was about fifteen when I started driving. I drove that old Magnetawan I guess fifteen or twenty years, for Graves Bigwood, then for Holt's. We used to drive from above Ahmic Harbour. Some of the logs were cut in the Township of Spence. We'd go up there on the first of May and come to Byng Inlet about the middle of November. We started one drive breaking ice at the Bell Settlement and ended up breaking ice in the fall at Byng Inlet. There'd be drives coming out of Gooseneck and out of Snakeskin, out of the Whitestone and Farm Creek into the Magnetawan. John Harris & Sons bought logs all around and dumped them in with Graves Bigwood's, and sorted them at Byng Inlet. A lot of them would go for one of those short drives then they'd go home and get their farming done, but I had nothing else to do.

ROY COCHRAN (b. 1905)
The first year I went on the river, I was fourteen in June of that year. The Still River. I drove the Magnetawan seven times. The logs came down Farm Creek into Deer Lake, and down the Magnetawan. Sometimes you'd get through fairly good, and sometimes, if you had head winds, you couldn't do nothing. Head winds pretty near broke Jack Campbell, and he was a good driver. The only thing you could do was watch, and whenever it calmed late in the evening or at break of day, pull the men out, and wherever logs had moved, pull the tail boom as far as you could with the gasboat, one end then the other. We took a gasboat all the way to Byng Inlet. It was built in Parry Sound by Croswells; they put a twin-cylinder motor in it. I think there were two portages we took it over. I ran the Canal Rapids. I wanted to run the Deer Lake Dam too, but they wouldn't let me.

That Canal Rapids is the most dangerous rapids. It has taken eight lives that I know of. I ran that rapids twice and came out of it, once with a gasboat and once with a punt with a bunch of boom chains in it. Mac Campbell ran it with me once, and another lad once. You watched, and wherever there were rocks you'd see the water boil up and you had to steer between them. The punt would be about sixteen feet long, a square-nosed punt, and the other was a pointer, a heavy log boat. It took a mighty good team of horses to pull it. We took a team right down the river with us, and I always looked after the horses.

Three Snye to Miner's Lake was a long portage. We loaded the boat on the wagon and took it over, a heavy thing to do. That road used to go clean through to Magnetawan [via] Island Lake, that was the halfway camp. Behind that camp, where they dumped salt [from salt-pork barrels], there was a hole eaten in the clay by the deer, big enough for a basement. The cadgeroad was grown up, a lot of it, but I followed a lot of it past the upper end of Deer Lake toward the Cataract, and from Island Lake through to The Graves and around Miner's Lake to Byng Inlet. There had been a bridge across the Magnetawan near The Graves, but it was all rotted away.

The Mountain Rapids was a bad one, and a bad place to portage. We portaged everything over to take it on down to Three Snye, our next camp, and Jack said, "Well, we'll have to portage the boat." I looked at the rapids, walked backwards and forwards a little bit, and I said, "No we won't, I'm going to run it." He said, "You can't run it, it's too dangerous." I said, "I can, and I'm going to run it." He said, "If you run that rapids and anything happens, I don't want no blame put back on me." And I said, "If anything happens to me, it don't matter." I shoved the boat out and went up the river a little ways to where the entrance came in, turned the boat around and headed back down to the rapids. They were all standing there separated out with their pike poles ready to hook onto my clothes and pull me ashore. They were wonderful motors we had. You can't buy a motor today that would compare with them. I just idled along — the current of course was sending it fairly fast — till I got down close to jumping over the rapids, then I pulled the throttle wide open. There was a back throw in it, and a side throw at the same time. I guess it was five or six feet of a jump, and when that boat hit the water below she just cracked like a rifle, went right out of it and never took a pail of water. Running with that big propellor and the motor wide open, when she hit the water she just lifted it.

124

AT WORK: ALLIGATORS
AND RIVER HOGS

JIM CANNING (b. 1872)

River hogs, that's what they called us fellows. You know, there's not much difference between one fellow and another. The river men, the real river men, were all pretty good. Perhaps some of them were pretty wild when they went out, but when you got working with them, boy-oh-boy there were a lot more good men than bad. A bunch of hard-heads, lots of people would call them hard-heads. But that's where you get some of the whole-hearted men, right amongst those that were supposed to be tough. Yeah, river hogs.

GEORGE KNIGHT (b. 1884)
[Excerpted from "River Driving," SYLVA, Vol. 6, No. 1]

Every stream that had a small lake, or even a marsh at its head which could be dammed to hold water, would be utilized to get the logs down to the larger waters. The dams were usually built in the dry weather of late summer, when the water was at its lowest. The banks of the creeks would be cleaned out so that when the dam was opened in the spring the logs would float down without danger of jamming. Dam building was a highly specialized job and a good dam builder was an invaluable asset to any lumber company. They had no surveying instruments, but they could tell with uncanny accuracy just how high a dam was required to raise the requisite amount of water.

Preparations for river-driving started a week or ten days ahead of the expected date of break-up. A gang of drivers would be sent in to the nearest lumber camp. Here the pointers would be caulked and painted, oars and paddles made, and peavies, pike poles and boom chains overhauled. Each driver would usually fit up his own pike pole. I have known them to scour the bush for days looking for just the right cedar burl for a hand hold. This would be carefully shaped, polished and attached to the end of the pole with a long screw, and almost invariably a copper cent was used as a washer. Just why the cent was used in preference to anything else, I was never able to find out. Each driver would usually go into debt for a pair of handmade driving boots. He would then proceed to fill the soles with caulks to his own liking.

An old saying goes something like this: "There is a tide in the affairs of man which, taken at the flood, leads on to fortune." With the lumberman there was a tide which, taken at flood, got his logs out of the small lakes and creeks and into larger waters before the water lowered. If this were not accomplished at the flood, the water might run off and leave his logs high and dry. This would lead only to the poorhouse, as only logs delivered at the sawmill brought any return. Everything depended on seizing the proper moment and working from daylight to dark. The lumberman could not afford to go to work in the middle of the afternoon, he went before daylight and he worked until dark.

When the ice had gone out of the lake or marsh, the boom around the logs would be tightened and the logs brought to the dam. The dam would be opened and the suction thus created would draw the logs into the current, where river-drivers were stationed to see that the logs went into the sluiceway straight. Other drivers on the stream below would watch for jams and break up any that might start to form. If the logs had been drawn out and piled on the bank of a stream instead of onto the ice of a lake, then the first job would be to break these rollways down and get the logs into the water. This was a dangerous job. The rollway might be frozen in the centre and require dynamite to loosen the logs. On the other hand, if the logs were not frozen together the whole thing, composed of thousands of logs, might come down immediately when the key log was loosened. Many river-drivers have been killed under a mass of falling logs.

Building the foundation for a saving dam. — GEORGE KNIGHT

Guiding logs through a dam at the start of the drive in the Pickerel River headwaters.
— GEORGE KNIGHT

— GEORGE KNIGHT

Saving dams on Caribou Creek and Four Bears Creek viewed from downstream at low water.
— ROY SMITH

As all rivers running into Georgian Bay were running more or less west, a continued west wind would mean a much longer time getting the logs down to the Bay. I can remember camping for six weeks at Dollar's Dam on the Pickerel River trying to get a drive of logs out of the mouth of the Wolf River, which comes into the Pickerel about a mile above Dollar's Dam. The wind blew steadily from the west, and what should have taken a week to do took six weeks. Wind of course did not bother logs running downstream with the current, but when they came to a widening of the river or to a lake, the head wind would stop them dead. If a lake was not too wide the logs might be run through loose, using glance booms to keep them out of bays and going in the right direction. If the lake was large the logs might be made up into bags and towed across. About thirty sticks of boom timber would be chained together and hung at the mouth of the river, one leg on each side. When one bag was full a hand winch would pull one leg over to the other, bringing another bag into place as it did so. To move these bags across the lake, what came to be known as a horse capstan was used. A horse capstan consisted of a crib on which was affixed an upright spool or spindle. A crib would tow six or seven thirty-stick bags attached to each other in a long string.

The "alligator" was the most outstanding invention ever introduced into the job of river-driving. It consisted of a flat-bottomed barge with a steam engine and paddlewheels to propel it, and a steel drum in the bow holding about a mile of three-quarter-inch wire cable. It was of very shallow draft and, as the river-drivers would say, could float on a heavy dew. It could warp itself over portages by the use of rollers, a screw arrangement being used to keep the boiler level. It was equally useful for warping booms of logs across lakes, pulling logs off shore and running the crew to and from work. They were so powerful I have seen booms and logs pulled right over a rocky point for, although there were always river-drivers on the blocks to keep them clear, they would get caught at times.

In the early days of driving, breaking jams and rollways was accomplished by the drivers entirely by hand. In later years it was done by a jam dog and decking line, to which a team would be hitched. If horses were not available a "Spanish windlass" might be used. Notches would be cut in two trees the proper distance apart. A log of ten inches diameter would be held in the notches while the decking line was attached, then the roller would be turned with peavies, winding in the chain.

The river-drivers were a hard-working, hard-drinking and hard-fighting crew, but withal, when they were sober, were generous, kindly and good natured. They worked long hours when wind and weather were favourable and were often wet for days on end. All they would remove when they rolled into their blankets were their caulk boots. If the caulks hadn't a habit of getting tangled up in the blankets, they probably would not have removed them.

River-drivers did their work when the blackflies and mosquitoes were at their worst. In order to get any sleep at all they would take a double blanket and sew up the end and one side, making a sort of tent. They would drive a stake at the head of their brush bed and one at the foot, and over this the blanket tent would be stretched, putting a spreader across the centre to keep the sides apart. If they stayed inside this contraption they slowly roasted, or probably stewed would be a better word, while if they poked their head out to get a breath of air, mosquitoes would be on them in thousands. I have many a time seen the cook at his bake board with a smudge pail between his feet and completely enveloped in smoke. If we found a few blackflies and mosquitoes baked in the bread, we just took them as a matter of course.

While the river-drivers worked for log hours at times, there were other times when they were idle for long periods waiting for a fair wind. At these times there were long sessions around the campfire at night when they vied with each other in telling tall tales. Many of the Paul Bunyan stories originated at these fires. To the river-drivers, the thousands of lakes which dot the North Country were but the hoof marks of Babe, Paul's blue ox. The Mississippi River was a leak from the water tank Paul used to make ice on his winter roads, and the griddle on which Paul's pancakes were baked was so large that in order to keep it properly greased, the bull cook was kept skating from one end to the other with a side of bacon strapped to each foot.

Al Cameron and Oliver Dixon breaking a jam below Dollar's Dam in the Pickerel River.
— GEORGE KNIGHT

Freeing a side jam on the Pickerel River.
— GEORGE KNIGHT

The Graves, Bigwood Company's alligator on Wahwashkesh Lake.

The alligator "Sweepstake" on the Pickerel River. — GEORGE KNIGHT

JIM McINTOSH (b. 1896)

When I was sixteen I worked on a sidewheel alligator, the *Nighthawk*. The logs went down the Black River into St. Jean Lake. There was an endless chain that went from there right clean over the road, over Graham's farm, and that put them in Lake Couchiching. We used to tow to the narrows at Orillia, then they'd cut the boom, and a tug would meet us there and take them down to Belle Ewart. They'd take down as high as 500 sticks of boom timber at one time. They were cut into logs down there, and some of them were shipped to England. They had to hew them on two sides to be legal. You couldn't ship a round stick of timber to England, it was a raw product. As long as there were two sides knocked off, it was a finished product.

An alligator was run by steam, a thing with big paddlewheels on each side. That's what drives it ahead. It ran on water and wood, good hardwood. We had different places to get wood. People along there got paid for cutting wood. Just let our rope go out full length, swing into shore, throw on a half cord or a cord, and away we'd go again. There was a crew of four of us — log man, a captain, a deckhand and an engineer. Old Cap Brooks only had one leg.

You have a bag of logs on 300 feet of rope. The reason you have the bag way back is so the current of the wheels won't hit the logs too hard and hold them back. You leave the bag hanging back 300 feet. If a head wind came up, the only thing you could do was go ashore and snub it till the wind went down. The wind on 10,000 logs, a forty-stick boom of logs, the alligator might hold her own, but she won't take her ahead.

The great old thing was the horse capsule. We warped down Big Trout Lake in Longford Township, Pickerel Lake, Bear Lake. The timbers [of the crib] were about a foot and a half square, hewed. It was snubbed to shore. A big drum in the centre, like a spool with a steel shaft down the centre. A rope went on that and away back to the boom of logs. An arm eight foot long went in there, and a horse got on the end and around he went, winding the rope on this drum. A man sat over there, coiling the rope as it wound in. The logs had to come. Have you any idea what a big horse, nineteen hundred, can draw on an eight-foot arm? He could go right through that three-inch rope any time he liked. When the logs were up to your crib, you'd take your crib away on down, and if you couldn't find a shore snub, a tree, you'd throw your two anchors out and warp the bag up to the crib again. Then lift your anchors, pull the crib down the length of your rope, throw your anchors and start winding again.

We used to warp at night mostly, unless the day was calm. That was one hell of a job on a cold night, a man sitting there coiling that wet rope. Somebody had to look after the slack, keep it coiled, you see, coil it on a corner of the crib. You could travel about two miles an hour, that's what they figured. That drum, I would say, would wind in about four feet of rope on every turn. Three men could get on that arm and walk it right around. When you got out on that bag of logs into the big part [of the lake] you had a punt rowed with oars, and you'd put your double anchor on the back of it and row away out there and drop it. Sometimes you'd throw the anchor quite a while before it caught something to hold. It was a hard job pulling in seventy-five or one hundred feet of rope with a 420-pound anchor on it. You had to lift them by hand. When the crib got up to it, two men would haul on it and haul on it an get her up over. They had a pulley on the front for the rope to run over. I've seen me at night, that rope would freeze when I was pulling it. You can imagine how nice that would be, sitting out there bare-handed and that wet rope.

I often sit here and think away back, the things we had to do then. Get out in the morning, maybe all night long, and it freezing. No use having mitts on, with wet rope and everything soaking wet. In the morning you'd think, Oh Lord, if I could get to a fire some place on shore and get dried out!

I drove the river for twenty-eight years, every river — the Magnetawan, Nipissing, Moon, Musquash. And I broke jams on Victoria High Falls, Raggedy, Dead Man Chute — mountains high — with a peavey. Start down at the front of her, get it loose so they'll all go. About three men at a time. In those days you'd hunt around a long time to find a poor man.

BURLEY HARRIS (b. 1883)

We used to be six months from the head of Shawanaga out into [Georgian] bay. There was ice on the water sometimes when we finished. We used to be figurin' on going to the camp again as quick as we could get home — knock around for a day or two, then back to the camp for the winter. Donald Campbell was the foreman on that river and I don't think anybody could beat him. He was the greatest man to handle water. Lots of them would have stuck the drive, but he knew how to handle water. If the river wasn't running right he'd shut her down and put in a dam or whatever it needed, and away would go the logs. There was a narrow gut that used to jam every time you'd fill it with logs. The company had improved the river the year before, and he had to raise every dam about three feet. That's where Charlie Croswell got his leg cut, improving that river. He got his knee cut and the first thing you know they had to take the leg off.

The logs that came down the Shawanaga went to Byng Inlet. They'd put storm booms on them, big dry pine booms, and they'd put great big chains in them. Put two sets of booms around a block in case one of them would get loose. Then they'd tow them to Byng Inlet. It was McCallum's logs that we were taking down. When we were driving out of Wolf Lake, McCallum came in with five or six greenhorns. They got out on the logs and there was the darndest pawing and swimming you ever saw. Old Mac jumped out on the logs with a pair of shoepacs on, no soles on them, and he cuffed around on top of the logs and helped fish the lads out. Some of them stayed and some of them didn't.

We had a jam on the Needle's Eye [on the Magnetawan River] one year and we had to let a man down on a rope to get the key [log]. He had a rope on him so that when it started they could take him up the rocks out of the road. He chopped away till she went "bang" and they jerked him up. They skinned him a bit. Then he had to go down again to hit it another crack or two. It was the first thing in the morning and the boss said, "You've done your day's work. You go to the camp and do what you like for the rest of the day." It was a pretty dangerous piece of work, cutting a log down in front of a jam like that.

GUY SMITH (b. 1885)

Sometimes you have to run water three or four hours before you start to run logs, because the logs travel faster than the water. It's got to fill the places where there's no water. If the logs catch up to your water you're beat. They'd pile up so bad the water would go on over or around them; they wouldn't lift out of there. Once it gets all bound up it won't move without dynamite or men to pick it loose. I drove the Still River. There are about twelve miles from the dam at Moose Lake to Britt, and I used to open the dam at two o'clock at night. We'd start to run the logs at six or seven o'clock in the morning, keep the water on till two in the afternoon, and they still had lots of water till dark. Handling the water was about the biggest job — it's the main job.

If a lake wasn't too wide, you could let the logs run through loose. Running logs loose meant you didn't use booms. Lakes that were big, you had to boom [the logs], tow them across the mouth of the river with a horse crib, cut them loose and run them through. They had a horse crib on Star Lake and a horse crib on Maple Lake and on Isabella Lake. They can't move them from one lake to another because they're big things.

A crib is made of big pine flattened enough for the horses to work on, maybe forty feet long and thirty feet wide. A little stable for a team of horses on one corner. The crib was good and solid. There's hardwood timbers six inches wide and mortised down six inches, goes right clean across. There's six or seven of them crossers, well spiked. Those cribs were hewed just as smooth as glass — good axemen at that time. Of course they've got to use shoes on the horses, because that pine would be slippery. The spool that winches the logs was right in the middle, sitting on a big iron rod that comes up through the pine. The spool is made of pine. There'd be a teamster and two or three men on a crib, then the boat crew, six men on the oars. You have a two-inch rope tied to the crib. The rope is coiled in the punt and you put your anchor on the back end of the boat and run it out the full length of the rope — maybe about 300 feet — and when you come to the end of your rope the anchor falls backwards off the end of the boat into the lake. They can't start warping till you get back with the punt, because the rope has to be coiled back into it [as it comes] off the spool. There are two arms that go out from the spool, one on each side, and there's a horse on this end and a horse on the other. The teamster could sit on the spool. When you get a good crib team, you don't even need a teamster. All you have to do is start them and they'll walk around. And sometimes they have to get right down and pull hard. You've got to put the rope around the spool about three times to get a grip. When the horses are warping, the rope works up the spool so far, then it surges itself back down, then up, then down again. There's a man who holds the slack in the rope, and sometimes he gets quite a jerk when it surges down. If it doesn't surge, just give it a little slack and it'll drop down to the bottom of the spool.

They maybe don't take the whole drive over in one pull. Sometimes it takes two or three pulls to take the drive across a lake. About three or four throws of the anchor and you'd have her across Star Lake. Wherever the water's deep it takes more rope and you don't get going so far. The anchor weighed 150 or 200 pounds. You cant warp with a head wind or a side wind. It's got to be calm or a fair wind. If it was fair wind the horses would step right out from the start; it took very little pulling to get everything tightened up. Maple Lake is about a mile across, and it would take about half the night. Do it mostly at night because it's calmed down. You get your block all rounded up and everything ready to take off as soon as the wind goes down.

Peter's and the Parry Sound Company drove the Seguin at the same time. One spring when I was on the drive for Peter's we didn't have much water, and the Parry Sound Company was right behind us. The way it was shaping up, the Parry Sound wanted to make a mixed drive of it to make sure they'd get their logs out. If they mixed, the two companies could use the water at the same time. The old fellow that was foreman for Peter's didn't want it mixed, and he kept a boom an the back of his drive so it couldn't mix. But they cut the boom and run into him, and the logs mixed. You had a lot more logs, but still the drive travels faster because you had more men, and there was always a gang moving ahead. Parry Sound Lumber Company would camp here, and Peter's would take the next camp. But then they had to sort the logs on Mill Lake. There's a small island near where the river comes in, and that's where we had the sorting jack. Look around that island and you'll see rock bolts all around it. When you're sorting logs you've got to sort with the wind. If you've got a north wind you have to snub on [the south] side so that the wind will blow the logs into your sorting jack. When the wind changes you've got to move to another rock bolt on the other side of the island, where you'd have a fair wind. You could move and sort at the same time; all you've got to do is change the snub from here to around the other side of the island and the wind will move the boom.

The logs were in a big bag, a circular boom snubbed to the island. The sorting jack is a floating outfit with maybe eight or ten men on it. There's a big flat boom, and there's a hole on both sides for the men to watch. You take the sorting jack out to the outside end, cut the boom, [fasten] one leg on that side of the sorting jack and the other on this side. The men watch the logs coming in and see whether they're Peter's or Parry Sound's. You have two bags hanging on behind the sorting outfit, one catching one fellow's logs and the other catching the other fellow's. Peter's stamp was the figure 5, and the cartwheel — a ring with a cross on it — was the Parry Sound. The logs would be hammered when they were skidding them; that was the roller's job. They'd tail a bunch down and stamp them, then deck them up. Peter's used an axe mark [also]. The chopper had to do that. Where he wants the tree sawed [into log lengths] he puts a mark, and right next to it he chops a V [the Roman numeral for 5] on that log and on this log. That's a log mark as well as the stamp. If they're sorting logs and the stamp isn't plain, they just roll the log and this axe mark is on the end.

They sorted all day, Peter's logs going into one bag and Parry Sound's into another. I had five or six men at night, cutting those bags off and taking them with a steamboat down to the dam. I'd snub one in a bay down there, and the other one I spilled at the dam. We had to have the boom timbers back up at the sorting jack by morning for them to go to work. When we got quite a few logs at the dam, we'd run them through into Georgian Bay then

Log-marking hammers, and the branded ends of salvaged sunken logs.

A boom of 12,000 logs off Good Cheer Island, Georgian Bay. — JOHN ANDRAS

sweep the river. Then we'd start to run the other company's logs down, give the other gang a feed of logs. We were all summer at it. There were enough logs to keep those mills going all summer. Lovely pine, nice big logs.

GEORGE BEAGAN (b. 1890)

I think the pronunciation of a river-driver was "a damn fool," because he didn't know when he was tired and he didn't know when he was hungry. All he knew was work. Work in the rain and sleep with your wet clothes on; that's the only way you could dry them. Take your boots off and sleep with your socks on. Make a brush bed — you have to use about four layers of brush, the coarse brush first then a little finer, till at last you stick the little branches straight up and down. You have a regular mattress then.

I think I had $120 when I came off the drive in 1906, and I banked $100 of it. That was a lot of money in those days. It makes me mad when I see in the papers that the fellows all went to the hotel and spent all their money. You take Burley Harris, Dan Kirkham, Eddie Vowells, Bob McEchern, Jack Campbell — I could name ninety and I don't think there was ten of them ever spent five cents in a hotel. Your dad was another, and Dave Little. They worked so hard for their money I don't think they wanted to spend it. When I came down [to Parry Sound] I heard them talking about ice cream. Mosleys had a place on Seguin Street. I went in thinking I could get five cents worth in a bag and walk out. I didn't know anything about ice cream in a dish. That's where I got my first ice cream, five cents. Then I had another dish and that's ten cents. I was only getting ten cents an hour. A whole hour's work.

MARSHALL DOBSON (b. 1891)

It was a little exciting when we used to move the steamboat, the old *Bear*, down over the high slide from Shawanaga Lake. We had to use three teams to take it around, and it generally took two or three days. She was a dandy old tug, all oak. The *Bear* started on Ahmic Lake and put a drive over there, and they brought it down and put four different drives on the Whitestone. It went down the river to Deer Lake — there'd be portages in places. Then they brought her out the road to Whitestone Lake and put her in again every spring. They moved her from there down to Shawanaga, and she put in three or four more years there. The last two years she went right through to Georgian Bay, and they brought her back up the road from Parry Sound on sleighs. Billy Craig ran her on Shawanaga Lake. They tailed the logs down the lake. Snub one end of a string of boom timbers to shore, hook the tug on the other end and sweep them along.

GEORGE DOBBS (b. 1905)

When they peeled the bark in the summer they cross-piled it right there. Then they'd take it away to a big pile, swamp it out to where they could get at it. We used a jumper, like a stone boat, then we drew it over to Ahmic Harbour with the teams. It was stacked there. We would swing the wagon around and hock her off the back and pile it on skids. Scows used to come to pick them up, and the *Wanita* and the *Mike* towed the scows up to Burks Falls.

Agar owned the boats. He had a hardware store in Burks Falls. He was a cranky, crusty old son-of-a-gun. He owned the *Armour* and the *Glenada* too. The Waltons started it first, then Agar cut in and beat them out. Knight Brothers' mill was downriver from Burks Falls. They used to raft logs out of Ahmic Lake, take them up through the locks. They'd take three or four rafts at a time, string of rafts. They had a floating camp with a rafting gang on it. The tug used to tow them around to these [log] dumps and roll them in and raft them up.

McCallum ran the *Bear* on Whitestone [Lake] for a few years, towing logs. It was built in an old barn [in Dunchurch]. It would be thirty or thirty-five feet long. Old Mr. Willard drew the machinery down from Burks Falls in the wintertime on sleighs. It had a Scotch Marine boiler, a return-tube boiler. The boiler laid down; most of them were upright. My dad rented it for a year and towed with it. Then McCallum was logging on Shawanaga, so they loaded the old *Bear* on log sleighs and drew her down to Shawanaga Lake and run her for a while. Then the next year, Dad wanted a boat, and that's when they started to build the

Dreadnaught in the old church that was across from the community hall. Dad ran it for three years before it burnt.

There used to be logs come down Staley's Creek and Eagle Creek, and they'd take the *Dreadnaught* up to the head of [Whitestone Lake] and use it right clean down to the dam. The water was six feet higher then; the roads used to be flooded before they'd get the drive down. They would kind of roll the logs through the narrows, anchor a boom and get a hold of the other leg and squeeze them through. When they'd get the logs down to the bridge [in Dunchurch], they had one section they used to take out. They would make arrangements what day — it would take a couple of hours. At that time of year there was always a hell of a current, and she'd go through to beat the band.

BERNARD MOULTON (b. 1894)

I drew logs with this alligator for five summers. Some called it the *Molly* and some called it the *Shoepac*. She had sidewheels. The Croft Lumber Company had it on Ahmic Lake. Their mill was two miles the other side of Ahmic Harbour. Their lumber went by scows to Burks Falls, then unload and take it to the [railway] station. They bought logs, and they had limits of their own in Spence [Township]. Before I started, Dave Bell was wheeling it. He tied up at the Croft mill one night and left some valve open. In the morning he went down and it was on the bottom filled with water. He never wheeled it after that.

It was a good job, wheeling that alligator, the best job I ever had. Tommy Dobbs ran the engine. In the morning we'd load the *Molly* with slabs — we burned dry slabs and you always had to wheel a bunch on. It would take about an hour to get steamed up, then take maybe fifty sticks of long boom timber up to where they were rafting. It wasn't very fast, slower than a walking speed with a tow, but empty it would go right along. They'd raft the hardwood. We'd hook onto the raft and bring it to the mill. Rafting, there'd be about eight of a crew. You rolled them in and put them in tiers, put a pole across them and dogged one end or the other. Birch will sink, you see. We'd leave McArthur's Bay up by Magnetawan with a raft after dinner and it would be about half past three when we'd get in to the Croft mill.

JIM McARTHUR (b. 1886)

Dodge's lost a quarter of a million dollars when fire went through about 1879. A big top fire, took all the pine. All the camps went, burned. They had just started to cut in most places — around the lakes and rivers — and dumped it straight on the ice. They hadn't gone far inland. They weren't going to spend a lot of money if they didn't have to. They had a forty-acre farm at the end of Miner's Lake. Graves and Bigwood grew oats and potatoes there. The cadgeroad from the Fourteen Foot, you could take a T Ford over the whole thing, it was so beautiful. It was cut out ten foot wide and there was a corduroy over a stream, like a bridge. Those lumberjacks could pick fine places for roads. They built the cadgeroads by hand. Mostly grub axes, a pick in one end and a blade in the other. Most of the corduroy, they adzed off the logs so they wouldn't be breaking stuff cadging in the supplies. The cadgeroad crossed just above the Mountain [Rapids], and went up into Sinclair Lake. A dam and a bridge on top of it; you'll see the cribs filled with stone there yet.

Where the graves are we call the Bridge Rapids, because there was an old original bridge there. The stringers were full length of the bridge. Pine. What pine there must have been! They laid them across, one tree each side. They had no hoists or nothing, all done by hand. That's twenty feet above water. Is the cribwork of that bridge still there? Oh, wonderful bridge, logs all adzed off and guard rails on each side. They had no engineering experience, but they were wonderful structures when you look back now. How they got those logs across nobody seemed to know, it was done so long ago.

BILL SCOTT (b. 1887)

In 1903 I drove from this end of Blackstone Lake to the Moon River for the Conger Lumber Company. All those fellows I drove with on that drive, they're all dead. The drive ran out of Blackstone into Crane, out of Crane into Little Blackstone River, and that empties over the

Moving camp on the Conger Lumber Company's drive, Blackstone River. The boat on the left carries the cookstove and tents. On shore on the right are a capstan and its heavy rope.
— PARRY SOUND PUBLIC LIBRARY

Drive gang on the Magnetawan River. Donald Campbell is on the extreme right, Dougald is seated in the boat, and his son Duncan is on the right in the background. — DAN CAMPBELL

Horseshoe Falls into the Moon River. Once we put the logs into the Moon we were done with them. They towed them with the big tug to Parry Sound. They were all in a boom, and there were storm booms outside the booms that were holding the logs. We had a million and a half feet, and I guess they took them in about two tows.

George Charlesly had the contract for this drive. He went $4,000 in the hole; he went so far in the hole that he never saw daylight. It was his own fault. He was as bull-headed a man as I ever saw. He ran out of everything. The last three weeks, all I had to eat were water whelps and leeks. A water whelp is just flour and water baked in the oven. We ate four times a day. The cook would cook them and we'd eat them right fresh. You couldn't get your teeth into them once they got cold.

When he got down into Little Blackstone, instead of booming off those bays and letting the logs go loose, why he put them into a boom and towed them across maybe five or six thousand at a time. That's the way he lost his money. We were weeks there that we shouldn't have been there at all.

You should have seen the tourists at Horseshoe Falls watching our logs go over. There was lots of water. A fellow had a stopwatch and counted the logs as they went over the falls. We put 1,500 over in fifteen minutes. There were logs there you could hardly turn over in the water, a thousand feet in them.

That's where we ate our last water whelp. The boss said to us, "Youse'll tail over the Horseshoe, move everything down to the landing, and I'll go down to the Moon River and try to get one of the Sweets to take us into Parry Sound." He went to Moon River with the canoe. We had four boats, and two of them we had to carry over. The other two were bigger and we let them go over. We had seventy-seven pairs of blankets and all the cookery riggin'. And there were peaveys, pike poles, axes, saws and everything; we got it all down there. It came dark and no George Charlesly. So the lads said, "We'll have to unwrap that tent and pitch it." We unrolled this big tent and put it up and we all got into it.

About twelve o'clock at night we heard a paddle in the water. Finally he got out of the canoe and came up to the tent, just like a dog you'd hit a kick. He pulled the curtain back and looked in, struck a match on his pants and peeked in. He saw a corner nobody was in and flopped down there. I waited for a few minutes to see if the older ones were going to say something. Not a word. Me and this Vankoughnett were the youngest two on the drive. I said, "Is that you George?" "Yah," he said. "Where you been all day?" He said, "Down at Moon River." I said, "Did you have something to eat George?" "Yah." "Lotsa drink too, eh, George? Well," I said, "there's a lantern there, George. You can light it and scratch out my time. I'm walking out of here." "Oh," he said, "you can't walk out of here!" I said, "That's what you think. When starvation starts looking you in the face, it's time to start doing something." And this Vankoughnett, he jumped up and said, "Here's a pig with the same snout, only a little dirtier. You can make out my time too." Then there was the cook, Harry Cole, he said, "Yeah, and me too." He said, "If Billy can make it out of here by walking, we'll follow him." Oh, he started to whine and cry. "I intended to send Joe Cornfield and Bob Moody to Parry Sound with the canoe tomorrow and get Syme to come out here with his tug and get us." I said, "I'm not waiting for Syme to come to get us!"

In the morning we struck out walking. I guess we got a couple of miles away when we heard a tug blow. It blew and blew. Some of them said, "Jesus, they're blowing for us. We should go back." I said, "Youse can go back if you want to, but I'm not. We're two miles away now, and maybe by the time we get back that tug will be gone, and where would we be?"

We hit Salmon Lake around six o'clock that night. There was a family there by the name of Rowlinson. I went in and said to Mrs. Rowlinson, "You couldn't give us a crust of bread or anything? We haven't had anything to eat since nine o'clock yesterday morning. We come from Moon River." She said, "Away down there? I can't give you nothing to eat, but I got lots of milk." She brung out four or five pans of milk, six-quart pans. "My husband has gone to Otter Lake to get a bag of flour. If you'd wait till he comes back I'll bake youse a scone." I said, "No, lady, we're glad to get that milk. We're going to head out for Otter Lake." We got about as far away as from here to the pavement and she came out and said, "Mr.

Scott, I'll tell you what to do. We got a brand-new cedar skiff here that belongs to a tourist, but I'll lend it to you to go down to Salmon Lake." She says, "You'll likely meet my husband on the portage, or wait on the landing at Otter Lake for him. He'll lend you the boat he has to go down Otter Lake."

We got about halfway over the portage and we saw him coming with this hundred pounds of flour on his back. There was a stump there just the right height for him to set the flour on then sit down alongside of it. We told him our story and he said, "God, Billy, come on back and the wife will bake youse a scone as soon as I get that flour home. We ran out of eatables today and I had to go to Otter Lake and get this flour." "No," I said, "We'll not go back, but your wife said we could have the boat you had to go down Otter Lake." "Sure," he said, "when can you bring it back?" I said, "I'll bring it back tomorrow." So he got up and struck off with his hundred pounds of flour, and we struck off to where his boat was. We got into the boat and came down to Otter Lake Narrows. I had an uncle who lived there, John Scott, and we went in and, Jesus, we pretty near ate him out of house and home.

JACK CHISHOLM (b. 1896)

In the winter they drew the logs to the lake, then when the ice went out in April they started their drive. There were rapids to contend with. The logs will pile up on the rapids and you'll have to get out above the waist in cold water, roll them off the rocks to get them going. Sometimes you had to use dynamite. You'd take fifteen to twenty sticks of dynamite and lash them to a pole, and put a fuse you can light from above the water in the centre one. You find a hole down among the logs and shove that down, then run and get under cover. When she goes there are logs that go twenty feet in the air, and some of them cut right in two.

If it's too dangerous to use dynamite, maybe blow a dam out or something, well you gotta use a peavey. There's a pike on the end of it, a square spike, and there's a hook on it like a canthook. You pick them out till you find the key log. The foreman's always on the job. Some fellows wouldn't go out on a jam because they were too nervous. They called for volunteers, somebody that was supple and could handle himself. You worked from daylight till dark, and sometimes nine and ten o'clock, to get a jam broke. And you gotta be watching. When they start to move, you want to be moving too. Get off them. Sometimes they'll go ten or twelve feet and bang, they'll stop again. Sometimes you break a jam with a team of horses and a swamp hook, a log dog they call it. It's got a ring in one end for the line. There's another ring big enough to get your hand in back near the hook for a trip line, so you can pull it back. The fellow down on the jam will hook onto a log, and they'll start up the horses and pull that one out. If that's not the key log you'll pull it back again. There are two men on the other side of the river, on the trip line, to pull it back again.

I learned to swim on the river. I fell in and my log got away from me, and it was, as the old saying goes, "either frog it or drown." I did some awful splashing, but I got to shore. Then Sherman Irwin taught me to swim. He'd take me down on Sunday, or sparetimes. We went into the river and he showed me how to swim. But it's surprising how many men that was on those rivers couldn't swim a stroke. But they were good on logs. If you can hang onto your log you can come ashore with it. That Sherman, he was one of the nicest men on logs I ever seen. He'd run that there Flat Rapids on a log, just him and a log and a pike poll. Not too big a log; the bigger the log, the harder to handle. When they start to roll they go too fast for you. You can go around a small log in half the distance you can a big one. He'd use his pike pole, hold water on one side, and if it was going the wrong way hold water on the other side to bring it back. You see, the water's going faster than the log is. You never stick your pole on the bottom. If you did you'd be off that log quick as scat.

ALEX GALIPEAU (b. circa 1908)

In 1928 I helped bring logs down from Chaudiere on the French to Callander with the alligator for J.B. Smith. I worked deckhand and log man. The deckhand got a little more wages. That was the *Woodchuck*, a sidewheeler. It burned slabs, some green, some dry, some inside, and some piled on the back.

At the mouth of the Sturgeon River they had a sorting gap, and all the logs of six or seven companies came down — Smith, George Gordon, Alex Gordon, Pearce, Coburn and Hettler. Some were on Temagami Lake, some on Sturgeon River and some on Jumping Caribou, but they all drove down the Sturgeon together then sorted them out. They all came into the sorting gap lengthwise, and you shoved them this way or that. There's a pocket, or "bag,"of booms for each, and the logs are all stamped on the end. After we had three bags, the alligator would take them out to a line of pilings, piers built with stones, and lash them together, three in a row, with chains. That was what you call a tow. Then the *Seagull* or the *Screamer* or the *Marco II*, whichever of the three belonged to that company, would come from Callander to pick up those logs.

Bob Turcotte was one of the best alligator captains. He'd worked on them practically all his life. There was the captain, the engineer, the fireman, the deckhand and two log men, that was the crew of the alligator. The captain looked after all the jobs, the overseer. The engineer, all he did was look after his engine, do the oiling and help load the wood on when it was low. The fireman was firing, and lots of times he helped the deckhand to hook on or push logs away when the sidewheeler was going through a bag of logs, so logs wouldn't slip endways into the sidewheels. That thing would go through logs like nobody's business. She'd go over logs and they'd come out the back. It was a flat bottom, you see. The deckhand was the one looking after the cable, hooking those bags and looking after the ropes, all the snubs. The two log men had their boat, a punt about sixteen feet. Their pike poles, peaveys and chains were in there. Go around and see that every chain was well hooked and wired.

Our alligator would go out about a mile, drop his anchor, back up and hook onto those three bags and pull them out. You didn't steer nothing when the alligator was pulling itself, you just followed the cable. There was no wheeling then, just when the sidewheels were going. You could curve around crooked or shallow places as long as there was no wind and you didn't pull too heavy. The cable sinks to the bottom into the sand and muck. There were speeds on the winch. The engineer looked after the winch and he would get the signal from the captain upstairs. He had a little bell. A jingling bell meant open her up, and she would wind when she was small. Once the winch gets big you'd slow down. If she went too fast, when it was big it would pull the logs under the boom.

They come with a boat with a motor on it from the big boat, bring the storm boom and pass over or under the cable that the alligator is pulling. They figure they've got so many booms to go around there, we'll say a hundred pieces. They pull it double like that, and this leg, maybe forty of them, is pulled off that side and this other leg out here, and they let go. The alligator is still pulling, and when it's far enough, maybe a mile ahead, he gives a "poot poot," and the *Screamer* comes with his big hook and hooks on. At the time the alligator lets go, the logs keep coming behind anyway. They are there with the yawl boat to take the legs of the big boom and come behind. They bring the two of them — sometimes they pass one another — around and lash them on with chain. That's their storm boom, lashed on the outside of the other booms that the three bags are in. The length of the other boomsticks would be twenty-five to forty feet. The storm booms are only sixteen to eighteen feet, but great big. If they were long, one end would sink in the big waves and there would be a long piece the waves would go right over the top. But with those, the logs couldn't go out. But some of them still did in a big wind.

ROY COCHRAN (b. 1905)

We used the horses for breaking jams, used a jam hook and a decking line, but not that much. We generally broke them with a peavey, and if they were too hard to break we used dynamite. It was hard when you would get a centre jam. It took a good man to wade out into the swift water to get to a centre jam. Some men could wade to their armpits in fast water and others couldn't wade up to their knees. I couldn't wade too deep; I could only go up to my middle and it would pick me off my feet. I was too short and not heavy enough to stay down. You used your peavey, put it upstream, pressed on it and waded, then moved your peavey quick and went again. It tended to split the water between you and it. They knew which guys were best at wading, and generally they would be the guys. A few wouldn't take a chance, but there were very few river drivers that wouldn't try.

The Holt Lumber Company's tug CAROLYN *aground. It now lies at the bottom of Wahwashkesh Lake.* — CHARLIE STILES

Steam tug towing logs in 30,000 Islands of Georgian Bay. The boom has been tied across at intervals to enable it to pass between the islands. — PARRY SOUND PUBLIC LIBRARY

That Canal Rapids, in about two days you'd run 50,000 logs through. It wasn't too often you'd get a bad jam in there, but you'd get little centre jams and that was the bad one to get out. I went down there and broke jams six or eight times, down over about seventy feet of a straight wall of rock, on a rope. It was a three-quarter-inch rope, tied around my middle. There were two men holding onto it; they could hold you because of the pull over the rock. I'd take a peavey and go down hanging onto the rope. If you wanted slack you jerked on the rope. When you got on that jam out there and that jam broke away, all you had to depend on was the rope. The guys that were holding the rope couldn't see you; there had to be a guy further up on the bank that could watch you. You waved your hand. If you were too far out, you might wind up in the water before they got you pulled up. But you couldn't get away; just grab the rope and hang onto it. The first jam we got there, there wasn't a guy would volunteer to go down but me. I went because I had no fear of nothing. All I had in my mind was that I was going to get that jam out of there. Even when I was a kid they'd come for miles for me to go to a barn-raising. I'd run a two-by-four scantling as high as this house. I climbed a forty-foot jammer one day to take the blocks off it. I got about five feet from the top and the goddamn jammer upset. I stayed with it till it got close to the ground, then jumped. Never got hurt a bit.

There was a jam below the CN [Railway bridge] and the boss told me not to go out on it because it was a dangerous place, but I went out at it anyway. Bob Ball from McKellar — he was a good river-driver, Bob was — he came along and came out on the jam. I had broke a lot of it loose and I yelled at him, "You get back to shore, that jam is going to go when I pull this log!" So he turned to go back, then stopped and looked to see what I was going to do. When the jam broke, a log came around and swept him off his feet and down he went. I could see him just spinning between the logs and the rock. I thought there would be nothing left of him. He went about the length of this house before he went over the jump into deeper water. The boys were down below, but they couldn't reach him. He just came up like that and back down again. I took out on the logs, and them going like a corker. I had to run as fast as I could or he'd be away down river in no time. I saw him again and yelled at them, "He's coming up, grab him!" They reached out with two pike poles and caught him in his clothes and pulled him out. He had one shoulder knocked out and he was jammed up bad. I couldn't see how he wasn't all ground to pieces. We were close to the railroad and we carried him to the track, flagged a freight and sent him to Parry Sound. Boy, he was a tough little man. Fight, he'd fight the devil. You wanted to be a mighty good man that could handle Bob Ball.

NORMAN CAMERON (b. 1894)

A dry-footed river-driver was no good. If your feet were dry you wouldn't go where you were supposed to go. If there was a bunch of small logs out there you wouldn't take a chance on running across them if you had low boots; you didn't want to get your feet wet. If your feet were already wet you wouldn't wait for a boomstick to jam over there and cause a jam for a quarter of a mile up the river. You'd get over there right away. Supposing you'd sink down a foot or so, it didn't matter. The sooner you got soaking wet right up to the armpits, the sooner they considered you a better river-driver. Then when they started running through the dam and a log would jam up on the rocks, if you were soaking wet and a good river-driver, you would wade out there with your peavey and pinch it off. So the rule was that the first thing you had to do in the morning, when you got your pike pole and your peavey to go down to whatever boat the foreman was going to put you in, you had to step in the water, fill your boots full of water first thing in the morning.

ALBERT SCOTT (b. 1895)

The Canal Rapids, that's a wild place. They never drove, only when the water was high and the lake was backed up full of water. There was one place that if a forty-foot boom of timber ran into a rock and swung around it would shut the whole thing off and cause a jam. So they made a practice of never cutting anything longer than thirty-six feet. They used to let a man down over there with a rope on him. Those rocks are pretty near straight up and down, and the bottom is a little bit cut away. There are three big stones down there, and that's where the logs used to jam. They put two long boomsticks out a little more than two feet apart, and a wooden roller with a gudgeon pin in each end, to let the rope run down. The roller was there to bring him up; he didn't need the roller to go down. They'd put the rope around his waist, take a half hitch under his arms and let him over. He had a hold of the rope with his hands, and he used his feet to keep himself away from the rocks. He took a peavey and a pike pole. There was a place where he could stand them up against a rock, and he could jump from it onto one of these three stones. A boomstick that would go in and jam, he'd take his peavey and run it under and get it out, give it a roll or two and let it go downstream. He took his meals when he went down there, or they'd bring a lunch pail and put a rope on it and let it down to him. He often stayed overnight. They'd cut brush off a tree, and [give him] a blanket, and there was a place down there where he'd sleep. A lot of men didn't mind going down, but they hated going up and down two and three times in one day. It's not a very damn pleasant place to go down. I never went down. I wouldn't go down there for nobody. I never was worth a damn up high.

RUPERT GREEN (b. circa 1905)
I worked on the Seguin drive of pine and hardwood for Mark Rogers. So many of the logs sank that we put a wooden bottom on the Seguin. We lived in tents. You know, there are two gangs of mosquitoes, a day shift and a night shift. To get any sleep we had to light a smudge in a pail, carry it in the tent, let it sit for a while, then take it out and shoo the smoke out with a blanket. Once you got to sleep the mosquitoes didn't bother so much.

JACK CAMPBELL (b. 1887)
I drove the Magnetawan and I drove the Seguin. I drove the Madawaska and the Shawanaga and the Naiscoot. A lot of the driving I did, I was boss on the job. In the first years I was driving with my father, but in late years, regardless of who I worked for, McGibbon Lumber Company or any of them, I was the boss on the job.

You go and look the river over, and where it needs booms you put them on, and you've got your camp set up with a gang of men in it. The logs are spilled into the different places, and you go ahead driving. In the early days there wasn't any power, not even gasoline boats. It was all "armstrong" boats and ropes, that's the way you worked. You could take five million feet down as cheap as you could take four million, because the rear is the biggest thing. On a river like the Magnetawan, when you let the logs go the water is high and they go into all those darn snyes. If I had left the damn driving out, I would have been rich. Sometimes I made a lot of money, but there were other times I lost more. The time the fellow cut the boom and spilled the logs down the Magnetawan River at the Noganosh Landing at the top of the Mountain [Rapids], I lost more than I made in ten years. The fellow that done it is dead now, poor fellow.

BERT CURRIE (b. circa 1890)
When I worked on the drive for Dick Robinson, the only chains we carried were for snubbing booms to shore. Boom timbers were linked together with withes. A maple sapling about twelve feet high would be bent over and twisted from the top till it was well withed, then it was cut off and both ends pushed in the auger hole. Then a wooden knockdown was driven in to wedge them tight. The knockdown was long enough that it could be driven out again with the back of an axe from underneath.

JIM CANNING (b. 1872)
I was eighteen when I first got a job running a gang of men. This was for Tim Sheehan. I was only eighteen and these fellows were all older than me, and I says, "Get somebody who has been driving a few years more than I have." And he said, "I want you."

Right as soon as the ice cleared they would be driving. On a little creek, they'd drive just as soon as the creek was open, drive with the freshet. And the drive would last sometimes to the middle of July, and sometimes they wouldn't get through at all. One year on the

Magnetawan they didn't get them through in time to get them across the Bay to the United States — an awful lot of logs were taken across there — because there was no insurance for towing them on the bay after a certain time. I think it was the twenty-first of September, the time the sun crosses the line.

I worked two or three summers on the Pickerel River. The Pickerel was a nice little river. A fellow got used to driving on a river; there were rivers that you liked to drive on. I went up there the spring after the hard winter, the year of the calamity, when there was no work for anybody. The Victoria Harbour took out square timber and logs, and that was about the only drive that went down there that year. The square timber was taken out for England, they loaded it in a ship like lumber. It was all trimmed up. A square timber is just as smooth as if it was cut by a saw, beautiful. They'd fall a tree and if it wasn't good enough for square timber they cut it into logs, and they had a log drive and a timber drive together. The square timber was always standing up on the corner among the logs. They could never get it so it would lay level in the water.

The Graves Bigwood Company came up from the old Dodge Company. Graves Bigwood had about twelve camps one time, and they had different log stamp for every camp. All the logs would be stamped, then the company would know camp so-and-so took out those logs. And if they happened to get their logs mixed with anybody [on the river] they could sort out the stamps. Graves Bigwood had this alligator on Deer Lake for twelve or fourteen years. I was foreman on the alligator; I guess maybe I wore it out. Another company had a drive coming out of the Whitestone River, about three and a half million feet. We were taking their logs across Deer Lake; they were giving [Graves Bigwood] so much a day for the gang and alligator. Ritter [the woods superintendent] said, "You better take it in two trips." I says, "What's the use of coming back? Let them put it all in." I knew I had taken more than three million feet for the Holt Lumber Company. I knew that a million feet didn't make any difference to the alligator. I noticed that Ritter was there that evening when we started to pull them out on the lake. He thought, now he's going to have trouble with that boom of logs. But I could pull — I didn't care how many logs was in the boom.

We had our alligator right through Ritter's Narrows and over to Bottrell's Narrows Saturday night. You couldn't pull the logs through there, you had to run them through. Sunday morning, about half an hour after we were usually getting up, up comes [the other company's crew]. They were camped down below, and they came with their boats, two pointers full of men, maybe twenty all told. Their boss says, "Well, you slept in this morning, Jimmy." I says, "Yes, we generally sleep in on Sunday morning." He waited around. He thought that when we got squared around we'd go to work. We had to open the boom and take the loose boom through with the alligator and hang it below the narrows, that was the next thing to do. He says, "Well?" "Oh," I says, "we don't work today. Where are these men going?" "What?" he says, "You don't work today?" "No, this is Sunday. The company I work for don't work on Sunday." "Lookit," he says, "I'll buy you the best suit of clothes that you can buy if you will let [the alligator] go." And I says, "What about the engineer down there, buy him a suit? And what about the fireman, buy him a suit too? No," I says, "I never worked black. I never took anything like that in my life, and it's no time to take one now. That old boat sits right there because we are told by our company not to work on Sunday!"

Well, he got on his high horse and went over and cut the boom and let the logs go through. He got a few sticks of boom timber and towed them down through with the pointers and shut off the last narrows, put a boom across in front or the logs would have went out into lower Deer Lake. There was a pretty good wind that day, so away went the logs, all three million feet. They went down and the narrows held them all. And here we were behind, with most of the boom timbers that should have been down below. It took us three days to get the alligator and enough boom timbers through. Now if he had waited till Monday, we would have had the alligator and all the logs through. We could have done it in one day. But we wouldn't do it on Sunday.

There was a place at the Fourteen Foot [Rapids], where the jump of the water came, that if a log was seventeen feet long it would be sure to catch crosswise, and the rest of them would come down and pick and pick. I used three parts of a box of dynamite putting a drive over there one time. The [drive following] behind me that year was Beasley, and he was shutting off the water. When we came over the rapids they just piled in there as big as this house. I hardly ever missed a shot. I got to know just where those logs were crossed down at the bottom, and I cut a long black alder, pretty straight, and put two sticks of dynamite on the end of it. You go out on the jam and put your stick down, and when you got away down there you'd feel the log and mark it with your pole. You'd bring it back up, go on shore, tie the two sticks of dynamite on the end, go back and put her down where you had it marked. I put all my logs over that way, because there were so many thousands in one jam. When we were over that rapids we were done with the dynamite, and I says to Jack Gibson, "Now you look after the dynamite. Hide it so you can hardly find it yourself." So he did. When we moved we couldn't find the dynamite.

April 14, 1890 [In Magnetawan] I met T. Sheehan, manager for the Moor L Co. and the inventor of the Great Alligator. — D.F. Macdonald

Horse capstan crib on the Pickerel River. — GEORGE KNIGHT

Drivers eating one of their two daytime meals on the river. Caulks gleam brightly in their boot soles. — GEORGE KNIGHT

FOOD AND CLOTHING

ARNOLD McDONALD (b. 1907)
We used to drive the Deer [Ferrie] River for Tudhope and Ludgate. We were up there before the ice was out of the lakes, and we slept in tents on a beaver hay or brush bed with no stoves. You crawled in there at night, a few old grey blankets, left your clothes on, and in the morning you were up before daylight and broke ice to wash. Old Bill McGhie was the cook. You had beans and salt pork, that long clear stuff. Rusty, saltpetre in it. It was from the spareribs down, off old boars and sows. Oh, tits on it an inch long, you could chew on one of them all day. And corn syrup. He baked bread, pretty good bread. He wasn't too clean, but you had no choice. There was a fire outside and poles you sat on and smoke blowing in your face. Tea. No coffee and bacon and eggs them times.

NORMAN CAMERON (b. 1894)
Dad used to drink a lot of whiskey. He made a lot of money; he was good at whatever he went at — a good logger, a good contractor, a good jobber — he made money at everything he did. And when he got done the first thing he had to do was have a little drink. He'd get drunk for two or three days, then he'd sober up and go for months and never take a drop. How he got to be a cook, the cook jumped the job one morning when they were driving the Pickerel, and they had nobody to get breakfast for the men. Mr. McFadden was foreman and he went to the men's camp to ask if there was anyone who would volunteer to get breakfast. Nobody would volunteer, and they were tented just a quarter of a mile from our place, so Dad says, "I'll get some kind of breakfast if there's anything there to eat." There was some bread left, and some bacon. He fried bacon and made toast and tea, and that's all they had for breakfast. I say bacon, but the bacon them days was salt pork, big long sides with 150 pounds in the side, salted and put in big boxes. They used to call it rattlesnake pork.

They sent a man to Trout Creek on horseback to hunt up a cook. He was away two days, and Dad would run over to our place to get a recipe for something from Mother. She'd write it out and he'd go to the tent and try to make it. One thing he wanted to make was barley soup. Mother told him to just put some barley in a pot and boil it up good, and that was all there was to it. But she didn't tell him how much barley to put in for the size of pot he had. Those old-time iron pots, they held about two gallons. So he was going to make two gallons of soup for quarter time, when the men came for lunch. He put the pot half full of barley and filled her up with water. It started to swell and boiled over on the stove, and he started dipping barley out into a bread pan. He had the bread pan filled with barley and the pot was still boiling over. He flung the bread pan, barley and all, out into the yard, and flung the pot as far as he could fling it. It's out in the pond yet. He had enough barley to make soup for the men for a month. The fellow was away two days and came back with no cook, so Dad stayed on as cook.

WALTER SCOTT (b. 1893)
Pretty near all the river-drivers, you'd see them with their pants stagged off. They'd cut them off because sometimes the caulks will catch in the bottom of your pants and trip you. You'd have them all ragged up anyway, so you might as well cut them off. They'd wear socks above the knee. And they mostly all wore straw hats.

ROY MACFIE (b. 1891)
[River-driving boots] give you a nice feeling, some way. I went out to the West one time and took my driving boots, and everybody out there thought they were a funny thing. There used to be great controversy between Taylor's boots and Tom Buchanan's boots. Taylor

made the boots around Parry Sound and Tom Buchanan made them at Dunchurch. All the drivers around there had either Buchanan or Taylor boots. Taylor's was a nicer-looking boot. [Buchanan's] had just a plain toe, and didn't seem to be quite as nice a boot.

MIKE GIROUX (b. 1888)

We worked long hours, daylight to dark. Four meals a day. They had a pretty good cook, a Frenchman. His helper was Six Nations Indian from down at Bala. The old fellow used to make nice raised bread. He carried a barrel to raise it. It had holes in it to let the air out. Put a lamp or lantern in the bottom to get the heat up. The stove had holes to put poles through; four men could carry it. Of course they'd take the lids off; we'd carry them in a box. But after we had horses, we'd use the wagon to move the camping outfit down. We put our clothes in a bag, a grain bag.

JACK McAULIFFE (b. 1901)

Hector McFadden was foreman. He was killed after a while, going down a hill with a wagonload of pigs, something happened and he was killed. I remember him saying to me, "Have you ever been on a river drive before?" I guess he had sized up my lack of skills and ignorance of what was happening. I said, "No, I haven't." He said, "Well, if you get out there running on the logs you're going to fall in a few times and the water is mighty cold. How would you like to stay in and be the choreboy and cookee for the cook?" Which I did. Being the choreboy, it was always a problem to dig up dry wood for the cook. I would carry it on my back for 400 yards. Being cookee I had to get up first in the morning. The men slept in three tents, six to a tent, and I had to build a fire in front of each, so when the foreman came along with his so-called whip, lashed it on the tent and said, "Up boys!" they had a warm fire to dress by.

But I didn't have that when I got up. It's pretty cold in the early days of April in tents. As we worked down the river — I guess Jack Lake goes into Wolf River — we ran into a big snowstorm and below freezing weather in mid-April. So cold that the boys in the tents found an old barn — I think there had been some settlers nearby — and they went over and dug a hole in the hay and that's where they slept. I slept with the cook. When I got out in the morning my shoes were frozen and I couldn't get them on. The next night I went to bed with my boots on. My caulked boots were the only footwear I had. We were river-drivers. We moved about once a week. Most of the logs were taken out by the settlers in the winter. They had their skidways on the bank of the river, and as we moved along we'd dump the logs in. We had two blankets. When we moved down we went in the bush and got a few balsam boughs and threw our blankets on it [for bedding].

The cook baked bread. He told me he had been on drives up in Quebec where he baked in the sand, even baked bread in the sand. I looked after getting the wood and water, and the washing of the dishes and the preparing of the potatoes — that was the only vegetable we had. Always prunes, always Beehive corn syrup, always molasses. They used to have pancakes in the logging camps, but not on the drive. Raisin pie. They had a couple of boards set out and the pots were all lined up. If you were one of the drivers you picked up your plate, a knife and a fork, your cup, and you went down the line and picked out what you wanted. There's your pot of beans; maybe if it was breakfast there'd be porridge, and salt pork fried and potatoes fried from the night before, and maybe pie lined up there, and prunes. I think we may have got some more supplies when we got down to the Loring area. I remember the many bags of potatoes, and the pork came in wooden boxes. I guess maybe 250-300 pounds to a box. When we came to a rapids we had to find a way to get everything over. The stove was quite large; it had to be carried over.

GEORGE DOBBS (b. 1905)

Art Buchanan started a shoe business [in Dunchurch], then he went into the store business and Tom [Buchanan] took the shoe business. Before the war he was busy, because he made caulked boots for the river-drivers. The river-drivers always came to Tom to get shoes; he made dandy good shoes. Instead of the sole being cut off even with the side of the boot, he put about a half-inch all the way around. He thought it was a good thing to keep from wearing out the side of the boot. They were mostly a ten-inch top, but he'd make anything you liked. He used to get his leather in great big sheets, and Mr. Moulton did too, to make harnesses. They nicknamed Mr. Moulton "Oak Tanned" because he used to get oak-tanned leather to make his harnesses. And he had another nickname, "By the Cats!" He used to say "by the cats." He made harnesses for Mark Taylor when he was logging. He used to charge about $75 for a set of harnesses.

DICK BREAR (b. circa 1900)

I liked driving boots that came just about to the ankle. A pair cost $8 or $10; we were making a dollar a day then. I used to get boots from a fellow in Magnetawan; the Magnetawan boot had a good name. That used to be good leather. I've worn the same boot three years driving; the last year I'd have to re-caulk them. I'll tell you what's wrong with footwear now. It's not tanned with tanbark. It's tanned with liquids that burn the best out of the leather.

It's hard on [boots], being in the water all the time. You had to make sure and get them a little on the big side, because lots of times you wouldn't have a dry sock in the morning to put on your foot. You get a wet sock and try to get your foot in a wet boot, and it's a pretty goddamn hard pull. I've seen us on the Magnetawan River, we wore wet clothes pretty near a week. And in the little hollow where my hip laid, I could have washed in it in the morning. That's straight goods. The tents were used for years and rolled up when they were damp, and they got holes in them so you'd think they had shot them with a shotgun.

Sometimes if it was raining real hard there'd be room [in the cook tent], but most of the time you sat there around a fire all humped up. Go in the cook tent and get what you want on your plate, then come out and sit by the fire. You went around this table and took what you wanted, and if you wanted you could go back in again. Eat four times a day. On the Magnetawan drive one of the Ross boys was dipping in this meat dish — it was bacon, I think, more grease than it was meat anyway — and he thought he would stir up the gravy with this big spoon they had. He dipped it in a couple of times and lifted it up, and Jesus Christ, all that he had was blackflies!

ALEX GALIPEAU (b. circa 1908)

You bought a new pair of [boots] with twelve-inch tops and the shoemaker would drive caulks in them for you. Short ones were no good. When you ran on the logs and one would sink, the others would come together and hurt your leg. Caulked boots are different, they're built for that. They've got the big wooden arch and a sewed-up tongue, and the sole was all pegged with wooden pegs. Your heel was high, something like a cowboy's boot. Fraser made the best caulked boots, but I didn't like them, I got my boots at the shoemaker in town. I got ten-inch shoepacs, got an extra sole on and got the shoemaker to put the caulks in. Then in the fall, when you would go in the camp, those caulks would be pretty worn. While you were driving the horses skidding, it was good, you could jump over the log and never fall off. You just wear the one pair of socks, summer and winter. If you have them too tight, that's when your feet get cold. When you're walking behind the horses skidding, you're not cold — you're jumping, you can't stay on one side. Next spring they were worn pretty near tight. Get that sole pulled off and put a new one on, and new caulks, and you've got another year's pair of boots. That's how you had to calculate in those days, 'cause you were only getting a dollar a day.

GEORGE BEAGAN (b. 1890)

I was the only guy in the outfit that had a canoe, a lovely Peterborough, sixteen bucks hard-earned money. But I got the job of fishing down on Partridge Lake, me and Sally's Jack Campbell. He did the fishing and I did the paddling. It'd only take us a couple of hours for us to get enough fish for sixty men. Throw the troll out and haul them in, pickerel and pike, pretty near a third of a canoe full. I don't know who had to clean them. We were too high-toned for that.

Drive camp at Sand Bay on the Magnetawan River around 1915. Remains of an old lumber camp are in the background. — DAN CAMPBELL

River drive crew, believed to be on the Magnetawan River. — JAMES T. EMERY

Cookery crib on the Pickerel River. The pike pole and peavey in the foreground probably belong to the young river driver. — GEORGE KNIGHT

River drive cook's tent and stove on the Pickerel River. — GEORGE KNIGHT

151

Tom Sheridan's shoe shop in Parry Sound, 1908. Note the sign advertising river boots. — PARRY SOUND PUBLIC LIBRARY

ERNIE CARLTON (b. 1891)

On the Shawanaga drive I've seen us come in off the river black dark at night, and old McCallum would be sitting up, whittling a stick. He'd have us up again at four o'clock in the morning and he'd be there, whittling a stick. I don't know when he slept. Somebody said he slept standing up against a tree, for he was very scared of rattlesnakes. But he gave a lot of people work. He fed good: beef, pork, potatoes, carrots. And May Farley and Mag Fitzer, who did the cooking at Number One Camp, were darn good cooks. But the butter, he used to get butter in those wooden kegs, and god, you couldn't eat it. Rancid. Herb Atherton, Sam Stewart and Johnny Jackson were good cooks; between the three of them I don't know which was the best. You couldn't go into the cookery but you'd have to have a piece of pie or a cookie and a cup of tea. There were lots of cooks who didn't want to give you a bite between meals. Up on the Still River I've seen us go for a week that we never put the caulked boots on, but Atherton made everybody eat three meals a day.

JACK CHISHOLM (b. 1896)

The grub was good if you had a good cook. The company supplied lots of supplies, but it all depended on the cook. On the drive the cook always cooked in a tent. When they moved you'd have your meal at this camping ground in the morning and maybe you'd be four or five miles down the river for supper. Of course your lunch was always brought to you in a boat. You were on the logs at daylight. There was lunch on the logs around nine o'clock and then your dinner was brought out, then your afternoon lunch, and you got your supper when you got in. There was a trail on both sides of the river, and you walked down the trail to wherever the camp was the next day. Our beds were built of brush, right on the ground. If they were going to move, they'd tell you at night, and you'd roll your blankets up in a bundle and tie it with a rope you carried in your turkey. Some of them moved by boats. Now on the Seguin River they had to do it all by wagon, there were so many rapids.

GUY SMITH (b. 1885)

My first pair of [driving] boots I bought in the store at home. I thought they'd hold caulks, but they wouldn't. Fairly heavy soles, too, but once they got soaked they got soft and the caulks turned off. [Thereafter] my boots were made to order. A shoemaker at Seguin Falls used to come in to the camp in the wintertime, measure your foot, bring the boots in to you and charge $7 or $8 a pair. Boy, those caulks stayed in; they stayed up nice and straight all the time. They'd break off before they'd bend. Boots only lasted me a year. You might be up to the end of July on the drive, running over rocks and boulders, and they got dull. Then in the fall, when you go in the bush, you use your caulked boots, and they're pretty well worn out. Sometimes you can have them caulked over again. Take them old caulks out, but you can't put the new caulks in the same place. You dodge them around a little different. They were driven in. I used to wear twelve-inch boots. A lot of fellows wore long boots right up below the knee, laced up the side a little bit. They could work all day on logs and keep dry, unless they were sweeping. Sweeping the river you'd get wet no matter what you had on. Getting them big logs off shore, you gotta wade out up to your knees before you get enough water to float them.

[At home] you'd have to take your caulked boots off outside. There was a hotel at Maple Lake — a good hotel too, maybe five or six rooms. People would sometimes have to stay overnight on account of the Tally Ho [stage] going across to meet the boat on Lake Joe. In the spring, when the drive was going through and they got a head wind, well, everybody used to go up to the hotel. Mrs. Sword — an aunt of my wife's — used to sit out and watch for fellows with caulked boots. Three or four would come along and some would have rubbers on, any old thing at all so they could go in. Some would have caulked boots on. She'd make them sit down and show her their boots, show if there were caulks in them or not. If they had caulks in them, they had to come off. And if he didn't want to take them off, she would put shingles down on the veranda, and he'd tramp into them and away he'd go into the hotel. After he got so drunk the shingles fell off, then she'd put him out.

One of the few remaining stumps on Carve Island in 1967.

154

SINGING, DANCING, CARVING, CLIPPING

JACK McAULIFFE (b. 1901)

In the bush, liquor was never thought of; there was no liquor. Some settler in the Loring area gave us a dance [during the drive]. He may have had a booyaw and charged us fifty cents each. My god, we must have chopped up that floor that night. We square-danced with our caulked boots on. They had a fiddle and somebody called off. And enough women maybe for two sets.

WILLIAM JOHN MOORE (b. circa 1875)

When I was young I could dance, call off and talk to the pretty girls all at once. The men all worked away in lumber camps all winter, so when they came home in the spring they had wood bees, and there would be a dance at night. I used to take the team and round up all the women for it. Of course I never had to take them home.

JIM McARTHUR (b. 1886)

Carve Island, above the Stovepipe [Rapids], was a wonderful place. Them river-drivers would have made wonderful woodcarvers. The stuff they had on there — a lot of it was taken, you know. There were things like pike poles, peaveys, canthooks, of different kinds. They were all there, out of wood. Every time they came there were so many fellows on a Sunday, or when they were being held up by wind, they would sit there and carve. That was their outlet. But the same all along — look at the graves, a beautiful picket fence around [them].

JIM McINTOSH (b. 1896)

I won two prizes at the Exhibition for waltzing, with a lovely partner, my wife. She could really waltz. Away back in the country and no place to go, everybody could dance. My wife weighed 160 pounds and she could grab a bag of flour and throw it on her shoulder just like nothing. She could butcher a beef, run a threshing machine, haircut, dressmake, anything. Aw, I wish I didn't have this laryngitis, I'd sing...

[He sings]

> Come all you who drives on the river,
>   come listen to me for a while.
> And a terrible tale I will tell you,
>   of my comrade and chum, Johnny Doyle.
> We were camped on the Wild Mustard River,
>   at the head of the old Tamarac Dam.
> Just as soon as we'd eaten our breakfast,
>   we were out on the head of the jam.
> With the rushing and roaring of the water,
>   with our pike poles and peaveys to pry,
> never dreaming that one of our number
>   that day must so horribly die.
> Oh there never was better on the river
>   than my comrade, my chum Johnny Doyle.
> He had drove over it more times than any
>   just because he was reckless and wild.

But that day his luck was against him
　　when his foot it got caught in the jam.
And you know how the logs would be rolling,
　　with great flood from the Reservoir Dam.
Oh we picked and we pried for one hour,
　　the sweat from our brow it did pour.
When we pulled his poor body from in under,
　　it did not look like him no more.
He was crushed from his head to his shoulders,
　　and his flesh hung in tatters and strings.
And we buried him down by the hemlock,
　　where at night the whipoorwill sings.

　　That's called "The Wild Mustard River." It comes from the east country someplace. They say there are more graves on that river than any other. Years ago I could sing from morning to night, then sing all night. There's "Peter Emerly" ...
[He sings]:

My name is Peter Emerly, as you will understand,
　　I belong to Prince Edward Island, close by the ocean strand.
And in eighteen hundred and eighty-one when the flowers were in bloom,
　　I left my native country, my fortune to presume.
I landed in New Brunswick, that lumbering coun-tree,
　　and hired to work for a lumberman, as you will plainly see.
I hired to work in the lumber woods for to cut those tall pines down,
　　and while loading sleighs at the skids I received my fatal wound.
Now there's danger on that battlefield where the angry bullets fly,
　　there's danger in the lumber woods where the logs are piled high.
There's danger in the lumber woods, but death looks seldom there,
　　Until I fell a victim to that monstrous snare.
Bid adieu to my good mother, so gentle, kind and true,
　　she raised a son so soon to part as I left her tender care.
It was little did she think when I'd been but a boy,
　　what country I might roam to and the death that I might die.
But adieu Prince Edward Island, that garden o'er the sea,
　　no more I'll roam her flowered banks or breathe her summer's green.
No more I'll watch those gallant ships as they go sailing by,
　　with their coloured proud flags fluttering on that canvas high.

MARSHALL DOBSON (b. 1892)

There was one summer, it was three or four weeks we never done a — well, I cadged, but the gang never went out to work. All head wind. There was no use opening the dam. That's the year they hung the drive up. But everybody got his pay. We used to have war there. They used to initiate them for the river. If you went to the wrong tent, they'd take you in and pull so much of your eyebrows out, and your eyelashes, then cut off a wisp of hair. It didn't matter how close it was cut. That was put up on the tent pole with your name on it. Everett Farley, and I don't mind who the other son-of-a-gun was, but Farley was the axeman, they held me down against a log. I had long hair and they cut it off close to the scalp, in chunks. I got somebody else to cut the rest off that night. Jimmy Clelland always wore a whisker three or four inches long. He went to work one morning and they were laying for him, and by gol they took him in and cut his whiskers and put them on the string, and cut his hair and put it on. He says, "Curse your bones! If I get out of here I'll break every one of youse in two!" Jimmy had to promise to be good when they let him up. That was a rough place.

HOTELS

JIM McAVOY (b. circa 1870)

I had the old hotel in Whitestone for a few years [in the 1890s]. It was in a field where the road to Ardbeg crosses the Whitestone River. I bought it and the farm and stock from Hanlon for $1,000. The Dodge Company didn't seem to think the timber around there was worth taking off, and Hanlon pulled up stakes and went west with a lot of other settlers around there.

But two weeks after I bought the place, timber estimators for Graves Bigwood came in, and then the place boomed. My customers were mostly shantymen coming and going. One night I had 110 men stay there, mostly sleeping in the barns and on the dance hall floor.

Then I moved out to Dunchurch and ran the hotel there. A man named John Burns had it before. I had a livery stable and the stage to Parry Sound and Ahmic Harbour. I made some money speculating on horses and ended up with the Mansion House Hotel in Parry Sound.

June 11, 1897 Town is full of river-drivers. — D.F. Macdonald

ROY WAINWRIGHT (b. 1908)

My father [Sam Wainwright] was the last owner of the Whitestone Hotel. He bought it and a hundred acres in 1902 or 1903 from F.A. Syne, a taylor in Parry Sound, who bought it from Jim McAvoy, who saw the handwriting on the wall when the pine timber was gone and got rid of it. We got burnt out when I was four or five, about 1913. My dad wasn't much of a businessman. McAvoy could see ahead, that there was no farming, and once the timber was done, what was going to support the place? The tourists didn't go for that kind of life, standing up to a bar drinking whiskey and fighting out in the yard. The country was going dry anyway. Business kept going down, and Dad couldn't see that till it was too late. Then he did small-time sawmilling to keep the pot boiling.

We called them hotels, but those places were just glorified halfway houses, just a place with a roof to get a bite to eat and a bed. We could put up twenty-five or thirty people. There would be eight or ten rooms upstairs, with a stairs leading from the main entrance. The rooms opened on a long hall, and I remember there was a huge linen closet at the end of the hall. Then there was a great big room over the bar, full of bunk beds. A stairway led from the barroom up to that. We called it the ram pasture. It was for people coming through that had no place to stay, and to get in out of the storm they paid twenty-five cents for a bed. Most hotels set up pretty much camp style. There was one long table in the dining room — bowls of meat, potatoes and beans, and all kinds of cakes and pies, and the waitresses kept going around filling up the dishes. My mother did the cooking, but she had hired help. The hotel had a unique lighting system, it had carbide lights. There was a forty-gallon barrel of carbide with a valve to let in water and create a gas that was piped to the lights in the building. There was a cluster of lights over the dining table. We could stable about ten teams in the barns. I've seen teams tied up along the fence when it would happen that a bunch going each way would get conglomerated there. What they were after was food, and feed for their horses.

I remember loggers passing through, migrating from the camps. It was mostly Graves Bigwood and Holt Timber Company. There were people that boarded at the hotel from the time the drive went out till the sawmill started. They'd spring out there. There was Ed Bottrell, Ernie Carlton, old Chris Carlton, Duncan McRae and Jimmy Armstrong. Some of them had no homes. They'd spend their time sleeping and drinking. Some of them would drink all day, and you could buy a flask to take up to your room. A lot of them would just sleep and eat. Getting past their best, just glad to get some place quiet for a while.

It was a tough place to run a hotel. I remember hiding under something listening to the roaring and swearing and tearing going on. Lots of rowdyism. Maybe there would be a fight. I kept out of the road. There was no law about the hours of drinking. You came to the bar and you drank till you fell down. If the bartender wanted to stop you and you were bigger than he was, you didn't stop. They never sold much beer, it was mostly liquor shipped in in forty-five-gallon barrels. They drew it off with a wooden pump. You'd get a shot glass for ten cents. Then they had what they called a flask — we'd call it a mickey — and they had the bottled liquor, which was a better grade of liquor.

I don't remember this, but I remember them telling about it. Around those hotels there was always an old fellow hanging around looking after the barns and feeding the horses. They called him the barn boss. Our supplies were shipped in by rail to Ardbeg. In the fall they'd send a team out and bring in all the liquor and supplies for the winter. Soda biscuits were shipped in barrels, and barrels of apples, salt pork and whiskey. We had an old fellow working there by the name of Owen Young, a regular old character, and every time he went to Ardbeg for a load of supplies he'd come back drunk. There was no hotel in Ardbeg and there was no hotel between Whitestone and Ardbeg. Where did he get drunk? There was nothing missing. Finally some of the neighbours saw him going by sucking whiskey out of a barrel. He'd go to the barn and get two or three of the best oat straws, good-bodied oat straws, and he had a three or four inch nail. The whiskey barrels were painted, most of them blue, and they were stamped with burnt-in lettering, Gooderham & Worts or whatever it might be. This old fellow scraped the paint away and worked the nail down in one of these burn holes, and he'd get a straw in there and suck whiskey all the way from Ardbeg to Whitestone. Then he'd plug the hole and rub flaked paint in it, and you couldn't see it. Of course out of a fourty-five gallon barrel you wouldn't notice what one man drank. It was a conundrum for years how this old fellow was getting drunk.

There was a one-legged man begging from camp to camp. He'd go from one camp to another and stay overnight and take up a collection, then he'd come out to our hotel and stay there and drink up all his money. A lot of people were sore about this because he would pilfer from the working man then go and drink it. This time he got drunk and crawled up in the back of a democrat in one of the driving sheds that belonged to the hotel, with his peg leg sticking out the back. My brother, a teenager, and a young cousin were hanging around the hotel, and the two young buggers got a handsaw and cut about eight inches off the peg. What I remember is the cursing and the roaring and the yelling and the swearing when the old fellow woke up and his peg was short. He couldn't walk. It was a whole peg, shaped to fit the hip. My mother was so mad that these kids would do that. Even though the old fellow wasn't much good, she didn't think they should have done that to him. Around any hotel there was a blacksmith shop, and Dad was very handy. He had to make a ferrule and shrink it onto the peg and put pins in it, put it back together again.

BURLEY HARRIS (b. 1883)

Anybody that used to make a habit of follyin' the drive mostly drank, and just as quick as they got out in the spring they'd go right to the hotel and never come away. Maybe they'd have to borrow some money before they would get back to work. The Montgomery House and the Kipling on the harbour side [of Parry Sound] were the main places. A fellow I knew came down to Parry Sound after the drive, and he got full and went into a harness room at the Mansion House. They had a livery stable and the man that looked after the horses had a little office there that he slept in, and this fellow went in there and went to sleep. When he woke up he didn't have a damn cent on him, and he had over $200 on him when he went in there. Somebody had rolled him. I never bothered with the hotel. I couldn't fight and I couldn't run very fast, and I knew that if anybody knew you had money they'd take it away from you. A fellow used to bootleg at the Shawanaga saving dam. He was catching the river-drivers. [The police] would go up to catch him, but he had it buried out in the field and they never could find it. Some of the boys used to get drunk there. Say, we had the greatest times. Of course once we'd get down the river apiece, they'd never get back up.

April 14, 1890 The town Magnetawan is on a drunk, shantymen and river rats clawing around.
— D.F. Macdonald

158

THEY USED TO FIGURE ON LOSING ONE MAN A YEAR

ANDY AINSLIE (b. 1889)

They used to figure on losing one man a year on the Magnetawan drive. Three fellows drowned at the Burnt Chutes. There are two buried below the Needle's Eye, and one down at Deadman's Island in Deer Lake. Johnny Labrash is buried at the foot of Deer Lake dam, and there are two graves at the Canal. A French fellow drowned on the Whitestone River, and he's buried out here in the little cemetery. George Fleming was killed at the Fourteen. He did a fool trick, something anyone should know better than to do. A boom timber came crossways at the head of the Fourteen and they couldn't pry it, so he jumped onto it with an axe, cut it on the top side, swung over and hit it one belt on the other side, and away it went. And he went with it. He went down on a rock and it slid over him.

It was Morton and McDonald at the Canal. They drowned running the boats through. Billy McMillan and I ran the Canal one time, and there's a place about halfway down, if you're running a boat through, you'd swear it's going to hit this bluff of rock a hundred feet high. It'll go up within eight or ten feet of it and then it'll go straight sideways. A lot of them will turn their boat sideways trying to keep it from hitting the rocks, but if they let it go, why it will just throw itself out. That's how Morton and McDonald did it. There was another fellow in the boat with them, one or two on the oars, but they swam out. One of the men buried at the Needle's Eye is named McDonald too. Some say there were three McDonalds drowned, none related. There never was a McDonald who drove that river clean through to Byng Inlet. They'd come on there, and as soon as they'd hear about the McDonalds they'd quit.

BERNARD MOULTON (B. 1894)

My brother Earl went overseas in 1916, in the 162nd Battalion, and came back and drowned in the Magnetawan River. Charlie Baker was foreman. I was on the shore when they went over. I went out to Ahmic Harbour and broke the news. They didn't get him for seven weeks.

That morning we had been moving around to Maple Island from Ross's Rapids and a fellow by the name of Bill Shiers had come in and he had no caulked boots. So [Baker] said to Earl, "Instead of you going around with the horses, will you stay here today and let Shiers go with the team. You got caulked boots." Earl says, "All right." They came to Ross's Rapids and the boats had to go over a falls like about from the ceiling down. The logs had gone over and they were taking the boats down. The foreman said to Jack Courvoisier, "How did you do this in other years?" Jack said, "We ran the boats." Of course the water was never as high. Jack said, "I'll take one if you'll send a good man with me." So he turned around and said, "Will you go, Earl?" Earl never answered him, he just went and got in the boat. It was about eighteen feet long and pointed on one end, with oars. I said, "Earl, if you get in that boat, remember you're never coming out." Because I saw where it had to go over, and he didn't. I was down with the boss the day before and I saw where it had to go over. When the boat took the dip, Earl stepped over the seat. I never saw him after. The boat went end over end. Courvoisier grabbed the big ring on the back and never let go, he went end over end with the boat. It took ten men the next morning to pull the boat up along the shore, so you know how swift the water was. They were going to use the boat to look for him, but the water was too swift. They looked every day, went up and down the river where it was calm. There was no use looking for him in the swift water. We took boom timber and spread it across the river and put weighted netting down. When he did come up he floated down against the netting. But it was seven weeks after. He had a cut across the back of his head; I think the boat hit him.

I quit the day my brother drowned and never went back [to river-driving]. I came to Parry Sound and drove a coal wagon for William Beatty Company. That was in 1928, and I worked for Beatty for thirty-four years.

JIM McINTOSH (b. 1896)

I've seen men drowned. I've helped take them out. I've helped dig their graves on the bank. I helped take Billy Madgett out. He drowned at the foot of Miller's Falls. In those days there was no screaming and yelling and crying and go around making a great big fuss. Harve McLeish, he was drowned. His mother took him out of the river, and his wife was there. They just up and dug a hole at the foot of a hemlock tree and put up a plank. Oh, there were lots of them drowned. Go to the Victoria on the Black River today and I betcha you'd see the bones of people's legs sticking out of the bank where it's washing away. It's only a few years ago I saw the bones and boots and buttons of a man's pants laying in the sand there. Oh, they thought nothing of getting a man drowned. No mourning at all, no.

Every man that ever was drowned is buried under a hemlock tree. They choose a hemlock tree. Carve a board and here's what it said:

> Remember men as you pass by,
> You may be once the same as I.
> And though you may be stout and brave,
> You yet might meet a watery grave.

A great foreman drowned. Munroe. He drowned on a Sunday morning, and four of his men. [He sings]:

> It was on a Sunday morning in the springtime of the year,
>   when our logs piled mountain high and we could not keep them clear.
> Our foreman said, "Turn out, brave, boys, and with hearts devoid of fear,
>   we'll break the jam on Jerry's Rock, and for Egan's town we'll steer."
> Now some of them were willing, while some of them were not,
>   to work on jams on a Sunday morn, they did not think they ought.
> But then six of our Canadian boys did volunteer to go,
>   to break the jam on Jerry's Rock with our foreman, young Munroe.
> We had not pried very many off when we heard his clear voice say,
>   "I would have you boys be on your guard, for this jam might soon give way!"
> He'd scarcely spoke those words when the jam did break and go,
>   and it carried off those six brave youths and our foreman, young Munroe.
> When the rest of those brave shanty boys those tidings they did hear,
>   like lightning to the riverside their course did swiftly steer.
> They searched those roaring waters and they searched them high and low,
>   but crushed and mangled on the rocks lay our hero, young Munroe.
> They took him from his watery grave, brushed back his coal-black hair,
>   there was one fair form among them all and her cries rang in the air.
> There was one fair form among them all, a girl from Saginaw town,
>   whose screams and cries rang to the skies for her true love who'd been drowned.
> Fair Clara was that noble girl, that raftsman's true friend,
>   who with her widowed mother dear lived at the river's bend.
> The wages of her own true love the boss to her did pay,
>   and a liberal subscription was taken up by the shanty boys that day.
> Fair Clara did not long survive her true love that was drowned,
>   for before three weeks had passed away she also was laid down.
> She was laid to rest by the yard blessed, and also in that row,
>   for her last request was granted, to be laid by young Munroe.

Now come all you true born shanty boys, wherever you may be,
    there's two graves at that hemlock tree, and beneath that hemlock tree,
the inscription board with the day and date and also in that row,
    fair Clara and her hero, the faithful young Munroe.

That's a true song. It's over 150 years old. I talked to men forty years ago that saw that board. He was buried under a hemlock tree.

JIM McARTHUR (b. 1886)
Martin and McDonald were coming through the Bridge Rapids below the Canal in a pointer boat when they were drowned. McGhie, who was a clerk for Lance Little's father, was on that drive and he told me about it. From what he told me, that bunch had a Chinaman cook and he was drowned at the same time. There was a separate mound outside of the fence, off from the men in the plot. That was supposed to be where the Chinaman was buried. They surely picked a nice place to bury them. It's a nice sandy hill and there was an old maple tree hung down over it. It used to swing, and gee, it would make an awful sound at night! All the different kinds of sounds, and with the roar of the rapids you didn't want to stay there long. It was the same at the Three Snye, the cabin where the cook committed suicide and the blood was on the wall.

May 12, 1894 Thomas McGown told me that young Wilson drowned on Jim Ellis's drive at the Serpent Rapids. — D.F. Macdonald

MEL CAMERON (b. 1892)
I was going back into the camp for the Parry Sound Lumber Company, and I was to stay at Labrash's hotel in Loring. George Bruce was walking the floor, and I remember wondering what was wrong. I found out next morning his boy had drowned in the big rapids below Dollar's Dam. He'd run a boat instead of portaging it around. He'd told him, "Don't run that boat. Let the team take it around and put it in below the rapids." Well, he decided the dam was open, lots of water going, but he hit a rock, punched a hole in the boat and it sank. The other guy that was in it got out. I remember George saying, "Youse fellows might just as well have put a stone around that boy's neck and threw him in!" Because he couldn't swim.

ETHEL NORTH (b. circa 1887)
Ollie and Tom [Sands, her brothers] were river-drivers. They used to just delight in being out on the logs. They used to be on the Shawanaga and sometimes on the Magnetawan. Tommy went down the rapids at Knoepfli Inn. They never expected he'd come out alive. Got one finger cut off. He was up on a big jam of logs. Somebody had to go up, cut the boom and let the logs down, and he went up. The rest of them rushed down to the lower part of the rapids, didn't expect he'd come out alive. But he came out all right, and he jumped up and threw his mitt off like that, and his finger came off along with it. He wouldn't have been more than a young lad.

ARNOLD MADIGAN (b. circa 1910)
My father was Jim Madigan, James John Madigan. He was a bushman. He used to be a foreman in the bush and he used to drive a lot of the drives. People called him "John Dollar." He was on a drive on the Magnetawan. There was a man had died years before on the river drive and they had just buried him wrapped in a blanket. [Madigan] went to get dinner ready and he found the head, so he cut off a tree on the side of the bank and he hung this skull on it. He found a clay pipe and put it in his jaw. He really got some of the guys when they came to lunch.

GOWAN GORDON (b. 1909)
I was there the day Earl Moulton drowned. I was one of the ones that pushed the punt off at the head of the rapids. We were working for Tudhope and Ludgate. We were camped at

Ross's Rapids and we had finished far enough down the river that we were going to move to Maple Island the next day. Charlie Baker was foreman, and he came in at night and said he wanted a man to go to Ahmic Harbour and get Victor Neely, the fellow that had the cadge team, to come in the morning with his team for an extra team. Earl Moulton got up and said, "I want to go." So Baker said, "All right, you go and stay home all night and come in with Neely in the morning and load the wagons." Earl went, and his wife told us after that he woke up his two children and kissed them goodbye that morning before he left, which he had never done before. He came in and helped load the wagons, but instead of going around the road with them he walked down the shore till he caught us at Campbell's Rapids. Jack Courvoisier was a white-water man, always ran the punts, and when it came time Baker said, "You're going to run the punt, are you, Jack?" Jack said, "Yes, I'll run it, but I want a good man with me." Earl stepped up and said, "I'll go." Baker said, "Is he all right?" And Jack said, "Yes, he's an old hand on the river; that's fine." Three of us stayed at the head of the rapids and the other guys went down below to catch the punt with their pike poles at the foot. We pushed it off, but when it went down over the drop and came up on the boil, instead of shooting out it upended back up the river. Jack Courvoisier came up, caught at it, missed it, went down, and he caught it again and rode it the rest of the way through on his hands and knees — it was upside down. And Earl, I think it was five weeks before Earl came up. We were down on the sorting jack at Lovesick Rapids before they found Earl. Bernie Moulton [Earl's brother] was with us that day and he said, "I'll never drive the river again."

MARSHALL DOBSON (b. 1892)
I was on the river when Tom Fawcett's three boys drowned at the foot of Burpee Slide. They were sent to find the cows, and I guess they got on a log, tried to "river hog" it, and drowned in the eddy. I suppose the eldest one wanted to dip the others and he fell in with them. Bill Craig went down three times and brought one up each time. It was all right while he was doing it, but it bothered him after.

BERT LITTLE (b. circa 1890)
My father, Bill Little, was a foreman for Graves Bigwood. We lived on a farm on Manitowabing Lake, but he would only be home a few weeks in the summer. He died of pneumonia in the hotel at Dunchurch when he was fourty-seven years old. Ritter sent him up Whitestone Lake early in the spring to clean up some logs. Through being wet all the time, he got a cold, then pneumonia. One of the men was careless with his pike pole and put a hole in my father's driving boot. Then the horse crib was so water-logged the horses turning the capstan were splashing water on the men. To make it float higher he got some dry pine sticks, which they put under by all hands standing on them to sink them. So he got still more wet. He was fourteen days in a coma at the hotel. Near the end he woke and said to his brother, "Have we got all the logs out of Grassy Bay?"

DAN CAMPBELL (b. circa 1880)
One time my father was out near Orrville, what they call The Serpent, a bad place for coming over. There was a jam and he went to break it. The jam broke away and carried him down through. He kept tramping till he got near shore and caught a limb and pulled himself to shore. The rest thought he'd be drowned. Didn't drown, but got a bad clip on the side of his head. That'd be for the Parry Sound Company.

BOB McEACHERN (b. circa 1895)
When the drivers would go down the Magnetawan, they'd fix up the graves. There were three or four graves below the Canal Rapids. They were kept very nice. Their boots were hung on a crotched stick shoved in the ground. There was one of them fell down, and we put wire and a new stick on it. I never heard when the men were drowned, but the boots were pretty well dried out.

ALBERT BOTTRELL (b. circa 1905)

I remember [Graves Bigwood's] alligator. I remember the pine trees where they used to have snubs. They'd come down through the narrows — that used to be full of logs right down to our place — and that old alligator would just go right through them with its side paddlewheels. They had a big steel roller on the front and a half a mile of cable, and they hitched onto some big tree for an anchor. I remember I got lousy off one of the men on it. They were on the logs and came along wanting water out of the well. I went and got a cup for them to drink out of, and one fellow took me down and wrestled with me — I was about ten years old. He rubbed his head right into mine. I guess he did it on purpose. It wasn't long till I was crawling with head lice.

JIM McAMMOND (b. 1903)

My grandfather [William Crawford] was a schoolteacher, but he was keeping time for the gang [at the sawmill at Byng Inlet]. Him and this guy [William Hunt] were going down to Byng Inlet and he was drowned in the rapids at the top of the Canal. When they came to the top of this rapids, they went in to shore and [Hunt] was in the bow of the canoe. When he went to get out he made a fluke and shoved the bow out with his foot. My grandfather went into the rapids and he couldn't swim. He caught an old stump that was floating around in the eddy — this was the story [Hunt] gave anyway — and he floated around from nine o'clock in the morning till four o'clock in the afternoon, then let go and went down. [Hunt] said the last words he heard him say were "Lord have mercy on my wife and my two little girls, and my soul." He was only twenty-one.

ROY MACFIE (b. 1891)

Ollie Simpson saved me from drowning on Bolger Lake. The ice had just broken up and I thought I'd try to birl a log when nobody was looking. I had no pike pole for balance, and I fell in. The logs were half covered with ice and I couldn't climb out. Ollie Simpson heard me yell and ran the logs to where I was. He jumped on the log I was hanging onto and birled it so that it rolled me up on top. Then he built a fire to dry me out.

Ollie nearly got drowned too, when they were running skiffs through a dam on the Shawanaga River. He was in the bow and Billy Craig was in the stern. Going through the gate the skiff tipped up enough to catch on some piece of timber, and the bow was torn off. When it landed in the whirlpool below, it went under like a shovel. The stern swung to shore and Craig grabbed a tree and stepped off. But Ollie went down, and when he came up he went around in the eddy with people yelling and trying to reach him with pike poles. He disappeared and they thought he was gone, but after a while he came walking up the river bank. He had kept his head, and every time he came to the far side of the eddy he swam. He was quite a way down before he got himself out.

JOE McEWEN (b. circa 1892)

Doog Campbell used to listen to the Canal Rapids roaring at night, and he'd say, "It's roarin' for its feed of man." We didn't know whether to go back to work or not. That's where all the graves are, at the foot of the Canal Rapids. She was a bad place. Somebody had to be let down on a rope. Tom Sands, he said he'd go in without the rope. He was going to jump in. It's a hundred feet, you know. Of course they had whisky. Jim Sands, his brother, he'd take him away back in the bush, then they'd both start to run. Of course Tom had no more notion of jumping in than I had, and this time he was going past us and he said, "I'm afraid I'm going to make it this time!"

Ed Sands went on the Steidler Creek drive for McCallum, but he had never driven before. When he fell in he didn't know enough to go to the end of the log to climb back on. He just hung onto the middle with his head out of the water on one side and his feet sticking up on the other. Doog Campbell went out and asked Ed how he was. Ed said he was all right, but he wished he was out of there. Campbell looked across at Ed's feet and said, "The fellow over there must be in a lot worse shape than you," and he grabbed his feet and pulled him out feet first. When Ed went home somebody asked him how the job was. Ed said it was a real dandy job; you never had to bend over to get a drink.

TONY GREEN (b. circa 1900)
I drove the Magnetawan for Jack Campbell, taking a drive down for the McGibbon Lumber Company of Penetang. There are about two miles of rapids before you hit Miner's Lake. On this piece of river we had lots of trouble with log jams. One time my two brothers, Ross and Louie, and myself, and Ronny Land and Bob Ball were working on a jam of logs which started to move. We all ran for shore, but Bob didn't make it. A log popped up in front of him just long enough to delay him, so that when he did get ashore the bank was too steep. He jumped, but slid back down between the rock and the logs. When there's a log jam the water backs up behind, and when it starts it goes with great force. We ran along the bank and didn't see him for a couple of minutes, but finally he came up across a couple of logs and we pulled him ashore. He was badly bruised and scratched, but no bones broken. We made a stretcher out of poles and carried him about a mile to the railroad, flagged down a freight train and sent him to Parry Sound.

FRED COURVOISIER (b. 1905)
Over here at Fletcher's Rapids, Fletcher got drowned. He was a foreman. They were running boats. There are two rapids, and at the bottom of one there's a big stone that makes a big boil. They had six men in a boat, and they had boom chains and peaveys and all sorts of stuff to load it down. When they got in the boat one fellow told Billy Fletcher, "Stay out of the boat." Fletcher couldn't swim. He gave a cuss word and said, "I'll ride it." So he got in and they ran the rapids, but the boat hit that boil and upset. Four of them swam down through the other rapids, but Billy Fletcher and another fellow hung onto the boat. The boat went down to the head of the other rapids, and Fletcher figured he could jump it to shore. Just as he jumped the boat struck and the other fellow, they got him to shore. If Fletcher hadn't jumped he'd have been saved.

Jack Courvoisier had a close call in what they call The Fourteen, away down there. He went through a jam. He was picking a jam and went down in it, but all at once there came a gush of water and shot him right out in the open logs. That's all that saved him. He was down in the water in the jam, and the gush of water came up and threw him out onto the dry logs.

RACHEL IRWIN (b. circa 1910)
There was a jam of logs, at the Canal Rapids I believe, and Uncle Charlie [McGhie] and Billy Smith, being the lightest and quickest of foot, were sent out to break the jam. But before they found the key log, the jam broke and logs were flipping end over end like matchsticks. Uncle Charlie and Billy Smith went over with them. The rest of the men rushed along the shore to where they thought they would drift in. No one thought they would survive the bashing of the logs. Uncle Charlie was a marvellous swimmer and he made it, battered and bruised. Billy Smith made it too, but he had wrapped his arms and legs around a log and was battered unconscious. He had "frozen" to the log and they had a problem trying to get him free of it.

TOM KEATING (b. 1885)
At the Burnt Chutes on the Magnetawan River there's a hole, oh boy, you could throw a porcupine in there and never see it again. A sixty-foot boom timber would go in right out of sight. And there's a centre rock. One time four of us went out and cleaned a jam off it, rolled it all off, then they got the warping line on the punt and let it down in the current for us. The four of us got in the boat and never noticed that the rope went out in a circle with the current. When we let go of the island and the rope straightened, the boat was hanging over the rapids before the foreman yelled at them to pull us. The water was going past the side of that boat an inch from the top. I shiver yet sometimes.

Another time we had a boom [stuck] across [a rapids] and I took an axe and chopped in

164

on one side to the heart. Donald Campbell, the foreman, said, "No use in letting one man do it all. Give me that axe." He wheeled around and hit the other side, and it cracked like that! There were two French fellows standing on the rock, and one grabbed me. The log rolled out from under me and he hung on, dragged me up the rock. I never did know how [Campbell] did it, but he threw the axe from here to that house across the road, and as he swung the axe it threw him back and the other fellow grabbed his hand and hauled him up over the rock. There might have been more graves that time!

Yes, there are graves along the Magnetawan, quite a few. They used to get the daredevils and run the boats through the rapids. That's how some of them drowned. I saw two fellows go through the dam at Deer Lake on a log. They picked a log and dared one another and struck off. And the one, Bert Jones, couldn't swim. By god, that was dangerous. You can't tell what the log's going to do. Going through, all right, that's fine. The water's running smooth. But when you hit the hole down below, you don't know what's going to happen. It might go straight in and right over. They both had to leave the log they went through on, but the eddy below was full of logs. They might have been smashed all to pieces. Foolish thing.

June 20, 1891: They brought young Malone down from Mountain Portage, where he got hurt on the sawlog drive. He died at Capt. Stewart's today. — D.F. Macdonald

JACK McAULIFFE (b. 1901)

We had two boys drown at Bruce's Rapids, two of the boys from Quebec. This scared a lot of the boys and very few of those that started the drive stayed on after that. McFadden, the foreman, and a Cameron from Loring were going to take the motor launch over and they had these two big pointers with four oarsmen and paddlers front and back. They had these to take over. Apparently they never could get enough water and there was a side dam below Dollar's Dam on some side stream that they'd open up and a terrific volume of water went down. After lunch our foreman and one of our most skilled drivers and experienced paddlers took the motor launch over. They gave me the battery — there was a portage over the rocks — they didn't want the battery to get wet in case they took water. The foreman was anxious to get back; he figured it was too dangerous, the water was too swift. He wanted to get back and tell them not to take the other boats over. But there was a crew of six in one of the pointers and these two boys were anxious — there was only a need for two to go over in one. (This was the usual way to take them down a rapids, with your paddler in front and a paddler in the back.) The four oarsmen in a crew didn't normally go over. But these two boys were so anxious to get their ride over that they hurried with their lunch, and I can see them today, sitting in the boat waiting. They took the launch over and McFadden was on his way back to tell them not to go, but in the meantime they went. This boat with a crew of six got in the eddy or current some way and upset. The two older experienced drivers couldn't swim, but they were smart enough to crawl up on the boat. Two fairly good swimmers made the shore, and these two felt they could swim to shore. But the eddy carried them out and within twenty-five feet of shore they were drowned. You know, the whole crew could have climbed up on the boat. The two lads would have been better if they couldn't swim at all. It was the darn eddy; they didn't have far to swim, but they didn't make it. Early next morning they recovered the bodies and took them out by way of Lost Channel and Pakesley.

We got the bodies out and everybody was a little anxious and timid. Next day, down below Bruce's Rapids, it was time for lunch and McFadden said to me, "Jack, you pick a nice spot up there and make the tea for lunch." I went up about forty yards from the river bank and picked my spot, a nice mossy, dry area to build my fire and make the tea. I built the fire, then went down to the river bank to get the water, and when I came back I had two or three square rods of a fire. I yelled and shouted. McFadden, when he came up, was mad as hell. He said, "What in hell did you build it up there for? Why didn't you build it down close to the water?" "Well," I said, "You told me to pick a nice spot." He said I was going to have to pay. That really hurt me. You can imagine how I felt. I cried like a baby. The cook, he had the rifle there, and he was so goddamned mad he was going to put some bullets in the rifle. He was going to shoot McFadden. That evening I wouldn't even pick up a plate to go and have my dinner, and McFadden said, "Jack, aren't you going to have dinner?" I said, "No. When I get this fire paid for." He knew it wasn't going to do any harm, it only amounted to an acre and there was quite a bit of swamp around it. I started out. "Now look," McFadden said, "do you know your way out of here?" I said, "No, but I'll find it." He didn't fire me, I think he would have kept me on — it was one of those moments, you know. He says, "Now if you wait another couple of days, we'll be down to Squaw Lake and it'll be much easier for you to get out." "No," I said, "I'm going now."

It's quite a walk from where we were to Pickerel and I lost the trail. Fortunately what I followed was the old logging road that took me to Kidd's Landing. Schroeder Mills and Timber was driving ahead of us and they had permanent camps there. So I said, "Where am I?" "Where are you going?" I said, "I'm going to Pickerel." "Well," they said, "you're not on the trail. This is Kidd's Landing." They said, "When did you last have something to eat?" I said, "This morning." "Well," they said, "you must be hungry. Have something to eat with us, then you can go in the canoe with our clerk down to Pickerel to pick up the mail." Kidd's Landing down to Pickerel was about six miles. I was so delighted to get a ride. He gave me a paddle. I had some experience paddling a canoe, and I paddled like hell. I had nothing for blackflies. You couldn't go and buy something to rub on. My neck was raw from blackflies.

I went home, and one thing the drive did to me was send me back to school, with the money I'd saved on the river drive and the few dollars a month my mother gave me for working on the farm.

ROY WAINWRIGHT (b. 1908)
I'll never forget the day I decided never to river drive or bushwhack again. We'd got down to Miner's Lake with the drive when I got an ulcerated tooth. It got so bad my eye went shut. The old cook, his name was George Moyer, said, "Look, young fellow, you'd better get the heck out of here and get that tooth fixed or you're going to die in here." I walked out to the CN from the old farm at Miner's Lake, a couple or three miles. I didn't have any money, but they had give me what they called a company time order. You took this to their office in Parry Sound and they gave you your money. When I flagged the train at South Magnetawan the conductor that was on that day was a mean one, old George Fox. He wouldn't let me on the train without money. I told him I was in bad shape and had to get to town — he could tell by the look of me I was in bad shape — but "No money, you don't ride with me." There might have been a spotter on, I don't know, but as the train started I grabbed one of those gondola cars and got in. The bugger had a flat wheel on it. God, I held that [sore face]. The conductor caught me at Ardbeg and kicked me off, but the engineer, his name was Noble Anderson, said, "Come on up here, young fellow, and ride with me." He let me up in the engine and I got to Parry Sound. And I made up my mind, if that's the way I gotta make a living, I'm going to try something different. I thought to myself, that's the end of that for me. No more!

GEORGE BEAGAN (b. 1890)
After Graves and Bigwood finished on Steidler Creek, Arthur Macfie was taking out logs on a hundred acres he had, and he figured it could be driven down Steidler Creek. One evening we paddled up the creek about two and a half miles and portaged over a dam. Coming back we got brave and ran our canoe through the gate. Arthur's paddle got caught in one of the grooves that holds the stop-logs, and we upset. We were both good swimmers, so we followed the canoe to the bottom of the rapids. Arthur said, "That dam is not going to beat us!" So in the moonlight we carried the canoe up over the rocks, through the brush and trees, and ran through the second time without mishap. Us wet to the ears!

BILL SCOTT (b. 1887)
I drove five springs and I couldn't swim a stroke. But I've floated the hat. That was down on the Moon drive. When I was going down the last time, a boom chain caught me in the face, and I just knew enough to grab ahold of it. When I grabbed ahold of it I popped up, and

there was a fellow there on the boom. He reached down and pulled me up. I sat straddle of the boom I guess twenty-five minutes before I got up on my feet to go ashore.

April 20, 1877: Heard that P. Bolger was drowned on the Seguin River, G.L. [Guelph Lumber] Coy's drive. — D.F. Macdonald

JIM CANNING (b. 1872)

One year the Americans wanted long timber. When we got to Byng Inlet, boy, that was the worst job I ever had. We had to chain them all together, like we were chaining a crib. Put a chain through maybe fifty of them, that's the way they took them across Georgian Bay. I wasn't very good that summer. I was getting over a sickness that I never did survive. I never did survive the operation that I had when I was twenty-seven. I had dropsy. The doctor says, "Don't river drive, don't work in the woods, don't catch cold and don't drink liquor." He says, "I've seen cases like yours." Me thinking I was getting better.

BURLEY HARRIS (b. 1883)

There were lots of people that were a lot better on logs than I was, that could take a chain and run out to the end of a boom without a pole, drop the chain in [the hole in the end], give the boom a whirl and grab the other end. Good balance. Young Jake Knoepfli, I saw him trying to run the rapids down below Three Snye. He went through twice, but the last time he lost his log in the chute and they had to fish him ashore in the eddy below. Him and fellows by the name of King and Bennett were watching a boom of logs on Ahmic Lake and the ice went out that day. They took chains in a canoe and went out in a storm. They upset the canoe and all went swimming. The other two fellows got ashore some way or another, but Knoepfli couldn't make her. They took word back to his mother and she said, "I don't believe it. There ain't water enough in Ahmic Lake to drown Jake."

River driver running the loose. — PARRY SOUND PUBLIC LIBRARY

Dougald Campbell.

Dougald Campbell's home in Waubaunik. — R. CAMPBELL

IN THE FAMILY

NORMAN CAMERON (b. 1894)

There were lots of men that were dandy good workers but couldn't get out to a log in swift water. That's a trade by itself. My dad and Bill Shaughnessy, neither one of them were big men — the most they ever weighed would be 160 pounds — but they were considered the best river-drivers in the country. Dad could wade right to his neck in the rapids, and it just boiling. So could Bill Shaughnessy — stand right there and pry logs. Anybody else, away they'd go. There's a trick in it. They bend a leg and the weight of the swift water holds that leg down. My dad took the contract of taking a drive for the Pickerel River Lumber Co., taking it down the creek and putting it in [Jack's Lake]. He took the job for $75. And we had to run through one dam. My brother Mel was two years older than me. He thought he was getting to be a pretty good man, and he bought his first pair of caulked boots.

The logs went through fine, but when we went to put the boom timbers through the dam, the booms would stand on one end on the stones on the bottom, and there wasn't enough water to lift them. Dad got the peavey and went out and started to pinch them off. Dad was pinching on one side, and there were some stuck on the other side, and Mel started to wade out. He only got up to about here in the water, his feet went this way and he went down into the eddy. Then he thought he'd try a new stunt. He went back up, and the booms were standing like this [45° angle] on the dam, and he walked down one of the sticks, just got into the water to here and away he went. And Dad laughing. Mel never could get out to the boomstick. Dad waded out and stood in that water. He says, "That's the leg [on the upstream side] you're holding yourself with; the other one's just to keep you from upsetting. That one, give it a good shot up the river. The more water hits it, the solider you are to the bottom." He says, "That's how they made self-loading dams; the weight of the water holds them down." He'd wade out and you couldn't tell what shape his legs were in — he'd be up to the chest in water. He pinched those timbers all off, and that's the shape his legs were in. [Cameron demonstrated, standing with one side to the current, with the upstream leg braced outward at a sharp angle, knee slightly bent.]

BELLA DICKEY (b. 1888)

All my nine brothers were drivers, except Walter. They drove pretty near every river around here. They'd go in the spring till the driving was done, and they'd generally be home for harvesting. Then they'd go in the fall as soon as the bush would open, go to the camp and come home in the spring till time to go driving again. Charlie was put down an eighty-foot rock on a rope one time to break a jam on the Magnetawan River. Lije went through a rapids — it was on the Magnetawan too — and the only thing that saved him, he got his arms over two logs. He went through with his arms over two logs.

PEARL MacLENNAN (b. 1902)

It was twenty-two summers that my father [Robert Buchanan] drove the river, and he never learned to swim. Then he fell in and he was going down for the third time and he remembered somebody told him if you go down the third time you never come up. So he started to kick and splash and got to shore.

They'd follow the river all the way down to Byng Inlet looking after the logs. He'd come home and plough the ground for potatoes and Mother would put them in. They had one horse; they didn't put in much crop. Until they passed Deer Lake he'd get home sometimes. From Burnt Chutes was four miles, that was nothing for him to walk and go back in the morning.

He'd come home from the camps with these tricks. Grab ahold of the top of the door and put his chin up to the top. He'd have us all trying to do these tricks. Put a pin in the chair and go around the chair without putting his feet on the floor and pick it out. I could do that one. Lay on the seat and go around the back of the chair and pick it out with your mouth.

And don't touch the floor with your hands or feet. The broom trick, jumping over the broom handle between your two hands. Another one was to take the broom handle and stick a pin in the floor, from your elbow out to the end of your hand, put your knees and hands on the broom handle, reach out there and get the pin with your mouth. And another one was you stood on one leg and your other foot up behind you, and then you got down and up again without touching anything. Just with the twist of this leg you got back up. I think you could hold your toe. That's how we put in our evenings.

JACK CAMPBELL (b. 1887)

There were four of us Jack Campbell's, all related. My uncle, a brother of my father's, they called him John Campbell, knew him from the rest like that. He had a son Jack, they called him Curly Jack. Curly Jack was also Gentleman Jack; always had a walking stick, had rheumatism in his leg. Then there was Sally's Jack — Sally was his wife's name. And myself. My father [Dougald Campbell] was one of the best fathers in Canada. A very hard-working man, and he wanted everybody else to be the same, without any talking. He always said, "I'm a man of few words." You never heard him telling this man or that man to do a certain thing. You were put on a job and he expected you to know what you were put there for. He just walked along and saw you were doing the job, and that's the way he done things. All my life I've done the same thing, because I was watching my father.

He was born in Scotland. He was in his mother's arms when they left there. He was a logging contractor a lifetime. He had a great big family and he kept us all in pretty good shape. His father was a sailor. That's how he got up here, he came around the Great Lakes to the vicinity of Parry Sound. He came up here in the spring again and was waiting for things to open up, to go on the boat again, and they put on a big drive for [settling] Parry Sound District. Everybody started to get in on the ground floor. They put the Great North Road up through and everybody started to settle here. That's a hundred years ago.

Everybody that knew him would tell you that my father was a damn good fiddler, the greatest man for music. I've seen him, when he was an old man and we were working for Holt's, he'd be working in the bush every day and [playing every night]. He died at seventy-two, but that was just a bad heart. You wouldn't know to look at him but what he was only twenty.

MARION SCHELL (b. 1904)

My father [Peter Ramsay] was a lumberman, a jobber. He was all over the place in sailing ships, then he went back to Scotland and married my mother and brought her out to all the blackflies. He never made an awful lot of money. He had a piece of land at Dillon, but he wasn't a prosperous farmer. He didn't get any education; he came out to Snug Harbour from Scotland when he was nine and never went to school any more. But he was a fine writer — flourishes, you know — and a wonderful reader; he loved English history. He played the fiddle, he could make it talk. He was foreman in a camp when he was nineteen. He cut his foot bad and was laid up all winter, and his brother bought him a fiddle. And boy, he learned to play that fiddle. I was never allowed to touch it. It hung on the wall by the neck — there wasn't a case — and we weren't to touch it. I don't know how my brother Jim learned to play, but they found out he could play and got him one. Then when I was twelve I found out I could play. I could take that thing and no matter what tune I heard, I only needed to hear it once and I could play it by ear. So they sent to Eatons and got me one, a good one, $25.

I remember my father bringing logs up the road and dumping them in the Shebeshekong River. They had men from Penetang with French names. And me a young girl, a teenager, and all these drivers swarming around. Mercy, I didn't want anybody to speak to me, I didn't know what to answer. I was raised in a very secluded, backward condition. There was no road to Dillon then.

Every time the last log was going over the rapids at the bridge, my father took us all up to watch the last one go over. They'd take them out to the mouth of the river and take them to

Midland — that was Manley Chew — tow the bags of logs away to Midland. We lived about a mile from shore, and when the tug would come up they'd blow so many whistles and my father would know the tug was there. He'd be expecting them. He'd go out and see that they got away with these bags. I remember Manley Chew, a big stout man. He chewed gum in a little round — I've never seen it before or since — tin box with round cakes of gum. That's the gum he chewed and that's what he brought us kids. I was very small at the time and the gum was delicious. Many a meal he had at our house. My mother often didn't have much to eat in the house, but my father always said, "Don't worry, she'll have something on the table." She'd make a pan of hot biscuits and put up a real good meal with next to nothing, for Manley Chew. But he was a very down-to-earth fellow.

JIM McINTOSH (b. 1896)

My grandfather came from the banks of Inverness and settled down in Cooper's Falls. There was an awful dose of McIntoshes down through that country. Raising a barn, he had one arm taken off at the elbow and never saw a doctor. My father raised thirteen of us, seven boys and six girls, miles and miles from a railroad or doctor. My dear old mother died in Orillia hospital at ninety-three years of age. She had a record of bringing 411 children into this world as [midwife]. When my old dad was over seventy he could dance just like a boy. My mother and father's people were all Scottish, they all talked the Gaelic. My grandfather had an awful job to get away from the Gaelic. Nearly all the old Scots people could step-dance. You take years and years ago, you'd go to a dance and there'd be a lot of old people there. I've seen as high as six people up step-dancing at once, with somebody playing the violin. They'd all bring them wooden boots, and would them wooden boots ever clatter on a hardwood floor!

Log jam lower Magnetawan.

Logs on the Seguin River above the Parry Sound Lumber Company mill about 1906.

Skidding logs into the Seguin River from a bank dump in 1885. A forest fire has swept the area. — ONTARIO ARCHIVES

COMPANIES AND BOSSES

ROY WAINWRIGHT (b. 1908)

Mr. Ritter was a good man for the company. That's why they had him there. I was working on the river drive for a Graves Bigwood foreman by the name of Charlie Harris. We had taken a drive out of the Whitestone River, down into Deer Lake. On Deer Lake they had a steam tug for taking the logs across. The logs were sorted on that lake. Graves Bigwood and Holt both drove the same river, and sorted the logs there. Charlie Harris was a wonderful man to work for. He'd work the hell out of you if the wind was fair. As long as the wind was blowing fair you worked, and worked hard, but if the wind turned around and you couldn't drive, you stayed in the tent, you didn't have to do anything. One day Mr. Ritter came down to where we were camped. He talked away for a bit, then said, "Well Charlie, you're not making much headway today." Charlie says, "Yeah, we'll sure have to get going when the wind turns." "Well now," Mr. Ritter says, "there's some saws and axes there. You take your boys, go up the shore and pull up some of them deadheads and driftwood, and saw them into four-foot wood and pile it up for the tug." This is wrong. If you're going to work men night and day when it's good going, you can't work 'em when it's bad going. And Harris, being a good foreman, knew this. "Well, Mr. Ritter, that's a good idea, if I had some wood-cutters. I have a bunch of river-drivers here, but if you get me a bunch of wood-cutters, I'll put them to work." We never forgot that. When it turned fair, did he ever make time, and every man Jack of us was right behind him.

Another story goes — I was too young to remember it, but I remember the talk of it — Albert McCallum took a contract from the Holt Timber Company to drive the Magnetawan River from Knoepfli Dam down to Byng Inlet. On the way down he lost the water. A dry season, the dams didn't hold water, and he stuck part of the drive. And because he only got part of the drive out, the Holt Timber Company only wanted to pay him so much for each log he got out. They wanted to penalize him for every log he didn't get out. They counted the logs and there was something in the line of ten thousand logs stuck in the river. So the Holt Timber Company withheld $10,000 — in those days a huge amount of money. Then, lo and behold, in mid-summer there came a deluge of rain. McCallum had the dams all patched and the water came up, and that fall he got the rest of the logs out to Byng Inlet. He put in a bill to Holt Timber Company for $10,000, twice as much as the original contract was for. The court case went to England. McCallum, being no wishy-washy type, took his court case to England, and he won it. But he didn't have a cent left after. The Holt Timber Company broke him, lock, stock and barrel. Absorbed all his equipment and everything. He won the court case, but that was the end of McCallum. A case of big money against little money.

JIM CANNING (b. 1872)

The first I worked out was down on the Seguin River for Bill Little, foreman for S. & J. Armstrong in McKellar. Old Jim Ellis, he was a jobber for the Parry Sound Lumber Company. He had some square timber belonging to another company in the drive too. We had to wade through those trees in that scraw hole down by Taylor's [near Mill Lake] — the water was held up, you see. We had to drive the logs right out into Georgian Bay, and when we got down to Parry Sound, somebody says, "Are we gonna get our money? We better hang onto the logs." Charlie Sheridan and I were looking after the feed booms this day. Here were the chains and that was where we stood. Anybody that came there to loose the chains and run those logs [through the dam], we were to push him into the lake. That's what we got to do. So we did. Along came one of the head men, out on this flat feed boom. He came out there and reached down, and Charlie Sheridan put his pike pole in through between him and the other boom and lifted it, and he went backwards into the lake.

At that time our two foremen were talking to more officials. After this fellow got out of the lake and went down there, they waved their hands, "All right, let the logs go." The eight or ten of us there thought they had made some bargain with them for us to get our money and to let the logs go. So we let the logs go. Jim Ellis said, "Come on over to the hotel and get money." So we all went over to the hotel and got our money — fifty cents apiece. My gol! I thought, ain't that a dirty trick. There's a great big mill there, and there's a big boat out on the lake, and the miserable wages a man gets for wading up to here in mud for three weeks to get them logs out. I just thought, well, that's a lesson. I got my fifty cents, and it took fifty cents to treat the crowd. We all treated the crowd until our fifty cents were gone, and we were so merry we didn't care whether we got any more money or not.

WILLIAM McKEOWN (b. 1885)
I drove the Seguin River for the Peter Lumber company when I was about eighteen. I was a real greenhorn, I knew nothing about it. Started to drive up in Balsam and drove down Middle River and across Manitowabing, clean through to Mill Lake. Billy Foran was the boss, a good man to work for. Jack Macklaim was the tail boss, and there never was a finer driver. He knew all about it, and he'd do the work before he'd ask you to do it. Foran used to take me with him practically all the time, and sent me where he didn't want to go himself. It was early in the spring and the logs were all frozen together. The gang was up at Irwin's Pond, prying them off, getting them started, and he'd sent me up the shore to see what they were doing. They were all wading in cold water to the waist, but I never wet a foot all spring. Oh, he'd put me feeding dams. Lots of fellows, they'll jam in too many logs, make them jams purposely — you know, hold the drive back, make the job last and have fun. That was a trick a lot of fellows played, but I didn't know any better.

I worked for Jack Campbell at McCallum's camp on Steidler Creek. It was before the First World War and wages weren't very high, about $30 a month. The measles got in there and I came home and brought the measles to all the rest of the family.

It was quite a place for liquor. Liquor was cheap then and they had a man carrying liquor from McKellar, and nothing else. There was a terrible lot of liquor, that's the way it was run. But he got the logs out. No reason why he wouldn't, the timber was all up against the river. Short haul, where he did haul, where you couldn't skid to the river. I drove on Steidler Creek for Joe Farley. Started at Wolf Lake. We used to be walking up there hours before daylight, and lots of times hardly made camp before dark. At the last of the drive McCallum was always there yelling and roaring, coat off, always wore a white shirt, pushing everybody along. Gangs on both sides of the river pushed the logs in as the water went down, tailing down to get them through the dam. Water was getting short, you see. Dear knows how many of us were on. There were so many men there you might get something to eat at lunchtime and you might not. Bread and pork, that's all you'd get if you weren't right there. Wherever you go there's generally a gang ready to run and eat any fancy cookies or anything, eat 'em all. The lunch carrier came out a trail that crossed up by Darky's Dam, or it might have been Straw Dam, the first one above the camp. You had to cross on the logs there, and you had to get there before dark or you had a hard time, the logs were so small.

July 13, 1894: Old Peter's trying to make a deal with the Midland and North Shore L. Co. for the sawmill. July 14: Peter's bought the Red Mill and hired Bill Beatty to run it. — D.F. Macdonald

ON THE SEGUIN

GUY SMITH (b. 1885)
(Recorded at various points on the Seguin River)

Mouth of Leonard Creek

All the pine that was in that Boundary Lake country came down this creek. There was no brush then; lots of creek, lots of current. We ran right out into the lake, boomed them up and towed them to Blinco Chute with a raft with two horses on it, with a crib. It's all fast water between here and Boundary Lake; it wouldn't take the logs very long. There are two tumbling dams up there. A tumbling dam is a dam to raise the water, but with no stop-logs in it. They have to run the water for a while before they start to run the logs, to raise it level with the tumbling dam, then it'll carry the logs right over the rocks. It's about five miles from here up to Boundary Lake. We used to be through here and right into Mill Kae at Parry Sound about the last of July, then I used to come home and help them off with the hay.

Blinco Chute

There was a dam across here, raised Star Lake about seven feet to get water to drive the rest of the river. That would hold a lot of water back. We used to have a lot of trouble here — a man almost got drowned one day. The logs had to make a short turn down there, and they used to hit that wall of rock and pile up. They had a man watching that all the time, but sometimes he'd get beat. You had a boom that you could swing across to shut them off until you could get the jam out down there.

Duck Chute

We used to camp here on the drive. At that time the river-drivers had sixty-six feet of shore; both shores belonged to the river. They could camp anyplace they liked along the shore, build a dam anyplace they wanted, do anything else on shore. The dam was right at the top, and there was a [log] slide over these big boulders, then a boom on out there so the logs wouldn't go too far. Then they made up a raft and towed it to the Isabella Dam. There were about seven or eight stop-logs in the [Duck Chute] dam. The logs all went through one gate; the other was just for the river to drain off if they got a big flood.

They had two men on each side feeding them through with pike poles. There's a glance boom that goes back to rock bolts on the shore on each side, and there are about three men who work on each to keep logs flowing in. They'd be coming down pretty thick because there's always a current here, and you'd put them through as fast as you could. We ran the pointer boats down through the slide. There's a jump off, maybe four feet or so, but you'd be going that fast that you'd jump right out in the open water. Somebody'd get on the back end, maybe two, to hold it down so when it makes the jump that end can't go up.

Seguin Falls

The first lake you come to up there [upstream] you call Horn Lake. Then you come to Axe Lake and Bear Lake. That all comes down through here. You have an awful flood of water some springs, and some springs you don't have any water at all hardly. But you've got to get the logs out just the same. They'd have maybe sixty men on, and they'd have it lined wherever the bad places were. When the tail of the drive was here, we wouldn't know where the head was, it was just going, that's all. They'd have a boom down on Isabella Lake in case the logs took right off and got through. They'd wait till they had everything in Isabella Lake before they started running to Parry Sound. It would be great fun jumping on one of those logs and taking down through here. It used to save a lot of walking; if you were going that way, just jump on a log and sail right along. The hotel used to be a pretty rough place. The river-drivers would go there and have quite a howdy-do. Up here below Fry's Lake there was a place where, if there was a head wind, you couldn't work. Those were the days they put in down here at the hotel. There was a fair wind down here!

Highway Bridge Three Miles East of Orrville

There's a bad place down there I used to watch. They call it the White Horse, and it's a white horse all right. Every year I was sent there. If a boomstick comes along, he won't always make the turn, he'll run his nose in behind a rock and the other end will swing. That forms a dam, and you've got to cut him out of there. They sent maybe a hundred boomsticks down [with the logs] so they'd have them at Isabella Lake. They always had a few put away there; the last [drive] to get off Isabella Lake, they'd pull some into a bay, take the chains out of them and fasten them to shore ready for next spring.

I used dynamite. I'd wrap six or seven sticks on the end of a pole maybe six feet long, put a cap on the end of a fuse, put it in the dynamite, and have the fuse running up the pole five or six feet to give me a chance to get away. That fuse burns a foot a minute, so you have lots of time to get out of the road. Where the current's hitting the boom the hardest out in the middle of the river, that's where you put the dynamite. Let the pressure of the water hold it against the boom, and it will cut that boom right in two. At that time there were some hemlock trees there, and I peeled a few butt rings off them and made myself a little shack up on top of the rock and put some poles and brush on it. When it rained I was nice and dry and comfortable there. Every year I was sent to the White Horse.

On the McDougall Road

We'd always camp at this bridge. We used to tent right there. It was all cleared then. There were none of those trees hanging down over the river either, that stuff was all cut away and kept cleaned away from the river; nothing like that left to interfere with the logs. Those trees would all have to be cut before you could drive it again. We were in tents, a cookery tent and about four other tents. There would be around fifty men in a gang. The next campsite up there would be at the Serpent Rapids. When they moved from Isabella Lake to the Serpent, they moved by horse and wagon, came in on the north side. I drove the river team from the Serpent. They rented the horses from a farmer that would be on the drive every spring. He'd go and get the team, but he didn't want to drive. I said I didn't mind driving.

I had a wooden-wheeled wagon with a vee-hayrack on it. You could take the whole works in one load. Then you'd have to go back for the boats, because you can't run the boats down the Serpent.

Log drive probably at Burnside's Bridge on the Seguin River. — PARRY SOUND PUBLIC LIBRARY

176

Part Three:
WORKING IN THE MILL

THE MILLS

MIKE GIROUX (b. 1888)

The first company that came in was the Dodge. They built the Byng Inlet mill. [Then] Holland and Emery. Bigwood came in and bought Emery out. Then Graves came and bought out Holland. That's how it became Graves and Bigwood. When the mill burnt down in 1890, they got a double mill. The big mill cut all the big stuff, and this other side, that they called the circular [saw] mill, [cut] smaller timber.

I did everything in the mill. I was on the carriage and tailed the edger for a while. I was tailing the edger first, then they put me behind the bandsaw and I got $5 a day. I was getting $3 tailing the edger. The bandsaw would take the slabs off. Nobody could stay there because he had to get those slabs out. Squaring up the logs, he had to step on the kicker, and this arm would come down and throw them over. That's how I got deaf, from the noise from the bandsaws.

ROBERT HARDIE (b. 1895)

My father brought a pit saw from Scotland when he came out with his father in the sixties. They were cabinetmakers and had used this mill there. It was the only sawmill around here in McKellar Township [in the early settlement days]. They homesteaded east of Manitowabing Lake, and cut the covering for the first Hurdville bridge with it. It was all above ground level, logs built up maybe four feet, whatever was necessary for the man down in the pit. There were two metal dogs opposite to one another that held the log. It must have been uncomfortable for the man down in the pit, with the sawdust, although they claim he was the best off because the other fellow had the hard work. Altogether it was said to be hard work. Pete Burnett said that there was lumber in their house, across the river from our place, that was cut by that saw, and I don't doubt the lumber in our house was cut with it.

Several had ownership [of the water-powered sawmill at Hurdville]. Matt Leach operated it for some time, then John Sirr, then Alf and Jim Parton. They took my brother Jim in

Parry Sound's first sawmill, which drew its power from the Seguin River, as it appeared in 1876 when it had been in operation nearly twenty years. — PARRY SOUND PUB-LIC LIBRARY

William Beatty who, with his father and a brother, acquired Parry Sound's first sawmill from its founders W.M. and J.A. Gibson in 1863. Beatty figured prominently in all aspects of Parry Sound life in the town's formative years and is generally regarded as its founder. — PARRY SOUND PUBLIC LIBRARY

Anson Mills on Byng Inlet, named for Anson G.P. Dodge, an American lumberman, was built shortly before 1870. — ONTARIO ARCHIVES

CHANCERY SALE

OF THE HIGHLY VALUABLE AND IMPORTANT

STEAM SAW MILL

WITH THE EXTENSIVE

Privileges, Lands, Timber Limits and Licenses,

IN THE

DISTRICT OF PARRY SOUND,

KNOWN AS THE

'ANSON' OR 'DODGE MAGANETTAWAN' MILL

THE TIMBER LIMITS COMPRISING (*i. a.,*)

The entire Townships of Ferrie, Chapman, Ryerson, Spence and Croft, and the East=half of Wilson, containing 352 Square Miles.

◄━━ ◆ ● ◆ ━━►

In pursuance of the Order dated the 24th day of April, 1874, and the final order for sale dated the 15th day of January, 1875, of the Court of Chancery, made in a cause of DODGE v. BARNARD, there will be sold by

☞ PUBLIC AUCTION

With the approbation of JAMES R. COTTER, Esquire, Master at Barrie of the said Court, by

E. S. MEEKING, AUCTIONEER,

——AT THE——

BARRIE HOTEL, IN THE TOWN OF BARRIE,

AND COUNTY OF SIMCOE, ON

WEDNESDAY, THE 3RD DAY OF MARCH, 1875

At Twelve o'clock Noon, in One Lot, the following Properties :—

1. All that valuable mill location situate on or near Byng Inlet on the Maganettawan River on the north shore of Lake Huron, as shown on a plan by Thomas O. Bolger, P. L. S., dated 6th July, 1867, of record in the Crown Lands Department, and butted as follows :—commencing where a post has been planted at the water's edge of the north shore of the Maganettawan River at the rapids thereof, and at the south-east angle of the said mill location, thence north 4 deg. 30 min. west 45 chains more or less, to a post at the north-east angle of the said mill location ; thence west magnetically 31

The diary of Duncan Macdonald reveals that on February 21, 1875 he left Parry Sound on foot for Byng Inlet, posted a notice on the Anson sawmill at 4 p.m. the following day and completed his 100 mile round trip back to Parry Sound at midnight on the 23rd. Macdonald was then employed by the Provincial Crown Timber Department, so it is probable this was the notice he tacked on the door of Anson Dodge's mill. This was truly a sign of the times, a worldwide depression beginning in 1873 proved ruinous to the lumber trade.

— RISSAH MUNDY

as a third partner. Jim rebuilt it about 1910, and it was built twenty-five or more years before that. Jim sold out to my brother Andrew, and he sold it to George Magee. Magee operated it for a short time and sold it to Bernard and Walter Johnson [who operated it until about 1950].

There were two waterwheels, one alongside the other. The flume and bulkhead were built of lumber. They had to be well supported with iron rods, for the weight of water that was in it. The fellow that owns it now tore that all off, for no purpose at all as far as I could see, rendered it useless. Most of the equipment came from mills in Parry Sound, when they were through with it. I cut a million feet one year, but sometimes when the water was low we didn't have too much power to finish the cut. About 10,000 [board feet] a day, that was pretty near the limit.

GEORGE DOBBS (b. 1905)

The first steam engine that ever was in Dunchurch [powered George Kelcey's sawmill]. It was a little Watrous engine, made in Waterloo, with a cylinder only ten by ten inches. It had a twelve-foot bricked boiler, built in with stone. It had a circular saw and a bull chain for pulling logs out of the lake, just a chain you'd hook onto the log and you'd press a friction into a wheel and pull the log up. Mr. Kelcey built the mill and ran it, then got killed in it. A board came over the saw and hit him in the head and killed him.

ARNOLD McDONALD (b. 1907)

Holt's mill ran around 45,000 feet a day, in one shift. Forty-five thousand in ten hours on a circular saw is quite a bit of lumber. They had the fastest steam carriage I ever rode on. I used to have lots of spare time, an easy day, and I'd be up on the deck. The first day I got there I didn't lean enough and, holy jumpin', my feet went out from under me. I wasn't long wakening up.

Tudhope and Ludgate's were on Gooseneck Lake. They used to run two shifts and run around 40,000 feet in a day. And Lew Robertson, he cut for Graves and Bigwood on Gooseneck and cut about 40,000. Graves and Bigwood had the biggest mill in Ontario at Byng Inlet.

ED PLETZER (b. 1887)

There used to be a dose of men around Parry Sound when those mills were going. It was a pretty wild place, especially at night. There were three mills in Parry Sound at one time. The Peter Company mill cut over 125,000 feet [daily], the Parry Sound Company mill was close to 100,000, and the Conger mill would cut about 45,000. There were an awful lot of logs went down that Seguin River, and sometimes they'd get mixed up and they'd have to sort them out. They all had a different stamp on their logs. Just past where the road goes over the river to the harbour, they had men sorting out the logs for each mill.

ROY WAINWRIGHT (b. 1908)

I fired for Feighan's mill for about three years, on Ahmic Lake, then at McKellar, a traction engine. I liked that kind of work. That's why I left the bush, I liked machines and there were none of those in the bush. My responsibility was to keep steam up and keep the engine in operation. A good fireman starts the mill and has it running all ready to go into the first cut when the whistle blows at seven. Sometimes we had a nightwatchman, but if we didn't I had to get up at five o'clock. The boiler was always hot from the night before. Put in a damn good fire and go and have breakfast, and after breakfast she's all ready to stoke up and get going. It's the fireman's responsibility to keep the machine in operational condition, keep the crossheads tight and the piston packing in, and everything that needs to be done. When we moved I had to move that traction engine on the road. It burned slabs from the mill, hardwood slabs. It would burn green ones, but if the mill is working like hell, green wood is hard to keep going without some dry wood to ignite it. It's a hard job sometimes, it all depends on what you're cutting. I hated cutting basswood. The slabs don't burn worth a damn, the trim blocks don't burn and the sawdust won't burn.

GORDON WHITMELL (b. 1899)

In 1915 I drew lumber all fall from Dunchurch to Ahmic Harbour for Tommy Simpson. Hosick had a mill too, and there was quite a bunch of teams on. Two trips a day. Put twenty-four or twenty-five courses of inch lumber on the wagon. No box, just the bolsters. Pine and hemlock mostly. We unloaded ourselves, piled it at the edge of the lake just past the hotel. We had to put on our load too, but there were men to hand it down to you. In the hot weather we used to pull out at five o'clock in the morning, before the day got too hot. Sometimes we'd get back to Dunchurch before the school was called.

BILL LITTLE (b. 1899)

I went to Lost Channel in 1921. I was looking around for a job, and I was going to go out on the [Federal Department of Marine and Fisheries steamer] *Lambton*. I hired on in the morning, and that afternoon I was downtown. James Ludgate, the manager for Schroeder Mills, met me on the street. I had worked in the mill for Ludgate at Lorimer Lake when I was seventeen, and I worked for him at Ludgate on the CNR. He said, "Billy, I'd like you to go up to Pakesley and run the jitney for me." He gave me a ticket and said, "You go up tonight on Twenty-seven." I said, "I'll be there." So I didn't go out on the *Lambton*. A big storm came up on Lake Superior and the *Lambton* went down, and they all were lost. I knew a lot of them fellows.

Schroeder Mills and Timber Company bought [the sawmill at Lost Channel and surrounding timber limit] from Lauder, Spears and Howland. That mill would run about 2,000 logs a day. Two engines in there, fired by sawdust, double-cut bandsaws. They had electricity for lights, a steam generator. The lumber yard at Pakesley was half a mile long, and they say at one time they had eight million feet of lumber there, and at the mill they had three million feet. There was always something going on at Lost Channel. They had an Orange Hall there, and a school, a hospital and doctor, even a big water tank on the hill, and they had sewers in. Electric lights came on in the evening. We went on picnics in the summertime, and fishing, and a lot of entertainment in the hall, in the wintertime skating. The village had maybe 200 or 300 people. It took a hundred men or more to run that mill. I heard Mr. Armstrong — he was the boss of the mill — say one time, "I went down to Parry Sound on Saturday and hired fifty men for the mill. I started the mill on Monday morning and only ten men were left. Forty went out!" You see, they could come in there, and if they didn't like it they could walk the track out to Pakesley and catch the train, and he'd be short of men again.

My first job was on the jitney. It was a Ford truck made for the rails. It had a turntable underneath and I could turn it around myself. I used to meet the CPR train [at Pakesley] in the morning and take passengers in [to Lost Channel], take the mail, and some freight, but not much because the train had a car for it. Then I'd come out in the afternoon again to meet the other train. I had it one summer, then the next year, in the spring, the boghole about two miles from Lost Channel got very soft, so they had the train on this end for about eight miles, and I was on the other end with the jitney. I'd bring the passengers and things down and transport them over and onto the freight to go out to Pakesley. They had phone orders. If you had an order to get in a siding and let a train go by, that's where you went. One morning there were four or five passengers on the boat from Port Loring and it was late, so I waited on them. They said, "You take the siding at Cole's Siding and let the work train go. The ballast train is coming with five or six cars." I started out to meet him at the siding, and I came to a rock cut about a quarter of a mile from the siding, and black smoke was coming. I said to everybody, "Jump if you can, because we're going to have a head-on!" It was April and pretty warm, and I happened to have the curtains up. I went out the side, and when I hit the ground they were just about together. I ran because I knew the glass might cut a person to death. Jack Thomas got his leg broke and a girl got a pretty bad cut, and a lady too. They sent a handcar down and we got them on and took them back to Lost Channel. They kept them in the hospital awhile, then sent them out. The jitney was smashed to pieces and they never replaced it.

Engine number 39 hauling lumber past Camp Six on the Key Valley Railway in 1922.
— BILL LITTLE

*A Schroeder's Mills and Timber locomotive in 1922. Mark Nichols is the engineer
and Bill Little the fireman.* — HUBERT HARRIS

That first winter I went firing on the main line for Mark Nichols. At Camp Six they loaded logs on flatcars and we'd pull them to the mill and dump them in the lake. We'd have about eight cars of logs; they had them staked and chained over to hold them on. They had a switch running down by the lake at Lost Channel. We'd back the cars down and they'd roll them off with canthooks. And we'd haul lumber out to Pakesley.

ARNOLD McDONALD (b. 1907)

In the fall of 1923 I went to Holt's mill and worked on the gas pointer at the jackladder. They used most of the slabs to make lathe, and from the hopper where they dumped the waste from the lathe mill they had dump carts. You came in and tripped her with a short pike pole thing, and your cart was full, and you went and dumped. It was about a hundred feet from the top, dumping down into the lake. There was a fellow on the dump and he'd pull the tailgate out. You just kicked a button and it upset, and then you'd go back. A continual round of pleasure; when you got back the hopper was full again. Well, one time they didn't get back in, the team and cart went down over the dump into the lake. I was on the jackladder and saw them, so I ran up the jackladder and yelled in and told them, then ran down and got the gasboat and ran around there. They were thrashing in the water and still hitched to the cart. I couldn't get right in to where they were, so I took the pike pole and ran across the logs. One horse was underwater. The other horse had reared and come down on top of him and he was drowned. But I hung onto the other horse, hooked his halter and hung onto him, until they got a small boat in and saved him. They got him out and fired the man that was driving the team, and the dumper.

December 26, 1889: Blowing a gale, the heaviest storm for years. The large piles of the Conger L Co's lumber was carried away in the storm and landed at Parry Harbour. — *D.F. Macdonald*

JACK LAIRD (b. 1892)

The Conger, the Parry Sound and Peter's were the mills I remember [in Parry Sound]. The Conger was down where Anderson's boathouse is, the Parry Sound was [at the mouth of the Seguin River] and Peter's was in behind the Kipling Hotel. Mr. Peter was a big, flush Yankee; you could tell by the talk of him he was American. He built that house Mr. Fenn had, and before that he lived in a big house by the harbour dock. He didn't drive just an ordinary buggy, it was something like what they went to the Parliament Buildings in, a big fancy buggy. But I don't think he was a show. I tell you those Yankees took a bunch of timber out of this country.

LEO MADIGAN (b. 1910)

I never saw one thing wrong with Mark Taylor. Sure, he came and bought the timber [from farmers] $4 a thousand. He paid $8 when we dumped it in High Lake. But they'd run to him to sell it for $4 a thousand. And you could go to his mill and buy a thousand feet of hemlock for $10. It wouldn't be number one, but you could get a thousand feet of hemlock for a barn for only $10. He hung on for four or five years, then made a big profit. Well, they said he stole that timber. He didn't steal that timber. They were quite willing to sell it for $4, and at the time he bought it it wasn't worth any more. About four years' cut he had up here at High Lake. He hung on until 1937. Lumber took quite a jump. Then he made quite a bundle. But to me that's business. He could have lost his shirt.

I worked for Mark Taylor nine years. The bush in the wintertime and the mills in the summertime. That's where I learned a lesson, in the mill at Maple Island. I used to tail-saw the circular. I got $5 more a month than what the slab-sawyers did. My brother Arnold and Newton Cooper were running the slab-saw. They said to me, "We should be getting more money. Let's kick." I said, "All right." I was getting $5 more at the time and I kicked for an extra $5. I kicked and they never did. Taylor says, "I don't know whether I can pay it or not." (He didn't say he wouldn't pay it.) So I went down to town and started on carpentry work, and three days after, I met old Mark on the street. "Say," he says, "when are you going to get back to the mill?" I said, "Never. You wouldn't give me a raise." He said, "I didn't say I wouldn't." "Well," I said, "you didn't say you would." He said, "I'll give you the raise. I've tail-sawed three days myself!"

His mill was right across from Geordie Wager's on the Magnetawan River. That was in 1937, his last mill.

W.T. "DUB" LUNDY (b. circa 1910)

Mark Taylor would always talk as if he couldn't make a dollar at all. I drove team around Mark's mill in Dunchurch, and Gilbert Carlton was the one that loaded the slabs onto the wagon. The edgings were tied up in bundles and put on the wagon and piled. They'd sell them, draw them down Snakesskin Lake to [International] Siding in the wintertime and load them in boxcars. Anyway, Gilbert and me had a real worryin' match trying to keep up. Taylor had bought a team of horses at Burks Falls from Knight Brothers when they went out of business, and if you'd walk in front of them and spit or anything, they'd stand right up on their hind feet. Somebody had teased them. And they were crippled up. Taylor came along this day and asked me how I liked the team, and that was my chance to get at him. I said, "The team is like your men, they're wore out. The team's wore out and the men that's here, they're wore out too. One man trying to do two men's work." I said, "I want a man for between Gilbert Carlton and me, and if I don't get him I'm going to quit." He said, "There's no men around." I said, "I know where there's one, Freddie Thompson." He said, "He's only a boy." I said, "He'll do me. He's a good big lad and he can work." So he took him on. It was the first job he ever had, and he stayed there till the mill finished.

When that mill at Dunchurch was finished Mark had another mill at Trout Lake. I went in with the team and he put me drawing slabs with Bill Lake. Our wages were $28. I got a whimper about how much Harry Wye was getting — he was on the trimmer — and some more guys. Bill Lake had one team and I had the other. The mill was starting at seven, and Bill and me went out to the barn. I said to Bill, "How would you like five more dollars on the month, and five more dollars for the months that we didn't get it?" He said, "How are you going to do that?" I said, "Don't put the harness on the horses. Leave the harnesses where they are and follow me." He said, "Where?" I said, "Right down to that office." He said, "What's down there?" I said, "Taylor." He said, "Do you think he'll do it?" I said, "He'll either do it or I'll quit. I'm going to get as much money as the rest of them's getting." I started down, and I could see Taylor standing in the doorway. "Now," he says, "I know what it is." "No," I said, "you've no idea what it is. I'm going to tell you what it is. I'm talking for Bill Lake and myself, because he won't talk I've got to talk for him too." I said, "We want five more dollars on the month, and five for the months that we didn't get it. We want the same as them men's getting in the mill." He says, "You name one man that's getting that amount of money in the mill." I says, "Harry Wye, and I could name you many other ones, but I'm not going to take the time. You know what you're paying them. I want the same." Well, the whistle blew and in about twenty minutes the mill was plugged, and he had about fourteen men doing nothing. He said, "The mill's running!" I said, "The hell with that. They can run as much as they like." He stood there awhile, then said, "You get to that mill as fast as your legs can take you!" He never said if he'd give it to us or he wouldn't. But the next month we got that five, and the fives back.

LES WELLINGTON (b. 1907)

The Holt Timber Company of Oconto, Wisconsin, had a mill on Deer Lake about three miles in from Bolger. I went there in 1924 and was there about three years. I went up to Bolger and into the office and asked for the manager, Charlie Carter. He asked me what I wanted to do, and at the time I was quite interested in steam, so I said I'd like to get a job firing in the mill. He said, "We don't have firemen here, we use mostly sawdust, but whenever we are cutting basswood we need a fireman, so I'll keep you in mind. In the meantime I'll give you a job fixing the track." I was only on the section gang two days when they started to cut basswood, and I went into the boiler room with the engineer and was there about a month. Basswood sawdust wouldn't burn, it would put your fire out just like you'd thrown snow on it. But you could heat fifty percent with hemlock or hardwood sawdust, and half slabs. The sawdust came down a chute from the saw into the fire, and if it piled up too high you took a long poker and spread it out.

They used to cut 70,000 feet during the two shifts. Some of the lumber pilers didn't work at night, they just piled it on flatcars and moved it out in the morning, but there'd be twenty or twenty-five men on the day shift. The office was right at the mill, and a cookery, bunkhouse, machine shop and stables. I was in the mill about a month firing, then there was a job came open on the locomotive, and I was there right up until the mill was done. They lifted the track and cut the locomotive to pieces for scrap.

They used to load the lumber in boxcars and take it out [to the Canadian Northern Railway at Bolger] with the locomotive. The year before I went there it wouldn't be unusual to go down to Deer Lake and see a steamboat going across with a boxcar with Canadian Northern or Grand Trunk marked on it, on a scow. That car would be full of grain or hay, and they would bring it in and put it out onto the end of a pier and onto a scow. A steamboat would take it from there up to the Landing Camp. From there another locomotive took it into Camp Eight, Camp Nine or Camp Seven, or wherever it was going to. It was a regular gauge railway.

There's a trademark of our carelessness up there that you can see right now. Where those cars used to go out [onto the scow], there are two flatcars in the water. They were used to load lumber and slabs on the night shift, and in the daytime we'd bring them out and the day men would unload them. They were old Grand Trunk cars, and the air brakes didn't work on them. Where we put coal in the tender was about halfway down to that dock, then we had to go right down to the lake for water. We had taken a trip out to Bolger in the morning and those two cars were in our road. Because the brakes didn't work on them and the track went up and down, we were supposed to put boom chains around the knuckles of the cars. But we came along that day in a hurry to get coal, and Jim [the engineer] says, "We'll just take them two flatcars down with us." I says, "Do you want to put the chain on?" And he says, "No, we're not going right down to the dock. We're only going as far as the coal chute." About halfway down he said, "Les, them cars is gone." The couples slid over one another from the wave in the track. They got away on us and right out onto the dock. One dropped into the water and the other on top of it.

They figured the best way to get them out was to anchor a cable to a big yellow birch in front of the cookery, run it through a block and onto the coupling of the locomotive, and we'd pull them up on shore. We had lots of power to pull, and the one car came up over top of the other and was just nosing up out of the water when the big birch tree pulled down off the rock right into the water, taking about four feet of earth with it. Now we had nothing to anchor to, so we just undid the cables and left the cars there. And they're still there today.

George Holt would come up maybe twice a year and spend maybe two days looking around. He was a pretty inquisitive man. You could tell he wasn't just one of us when you saw him. He dressed — remember the blue shirts they used to have with the celluloid collars? Oh yes, he was a real gentleman. Old Mac McCallum had trouble with him. McCallum jobbed for him, and they had a lawsuit over it. Holt beat him. He took it to the Supreme Court and Holt beat him again. He took it to the Queen's Court in England and he beat Holt. But by that time McCallum had spent all the money he had.

July 13, 1894: Vessels in [Parry Sound] harbour loading for Toledo and Tonawanda. — D.F. Macdonald

These aerial views of the Graves, Bigwood mill at Byng Inlet, taken in mid-summer of 1922, show lumber-piling areas well filled and several booms of logs remaining to be cut. — THE AIR BOARD, OTTAWA

Lumber wagon horses at Graves, Bigwood's Inlet mill. The two-tiered tramway permitted the building of extra high lumber piles. — EVERETT KIRTON

Lumber-pilers at Graves, Bigwood's Byng Inlet mill, c.1907. — EVERETT KIRTON

Moving the flywheel into the Holland and Graves mill at Byng Inlet. Its size can be judged by the man's head in its centre. — JAMES T. EMERY

James Thissell Emery, to the left in the picture, was a partner in the firm which operated the huge mill at Byng Inlet in the 1890s. He was also an amateur photographer.
— JAMES T. EMERY

QUICK RICHES

DON MACFIE (b. 1921)

Mr. Newby had the hotel in Dunchurch, and he got the idea that a steam traction engine would draw several wagonloads of lumber to Ahmic Harbour. He ordered one, a big Case, and it finally arrived by train in Burks Falls. He and somebody else went up and got it unloaded, fired up, and struck out on the wrong road, going away out toward Horn Lake. After several days they were down to Magnetawan, where they decided it would be quicker to go to Ahmic Harbour by barge. It fell through the deck of the barge, and they were all the time of the trip getting it jacked up to drive off at Ahmic Harbour. But they finally made it home to Dunchurch.

Fair Day came soon after, and Newby thought what a fine show to drive it to the fair. On entering the grounds he started tooting the whistle, and what with the steam blowing, it scared all the horses and cattle. Henry Ferris, the president of the Agricultural Society, got mad and ordered it off the grounds. But Newby owned land beside the fairgrounds, so he moved there and put on his show. Attempts to drive it up the Grange Hill were useless. It was taken to Whitestone to run a sawmill, and this was not a success either.

Cliff McCarl had been out west for the harvest and came home full of stories about how these traction engines could pull seven loaded wagons across the prairie. I guess he worked on Jack Hosick, who never was at a loss for new ideas to get rich, none of which ever panned out. Jack also thought one of those engines, pulling three wagons of lumber, would make it for him quick. He ordered a McDonald. It arrived at Waubamik, and McCarl brought this dandy new steam traction engine up the road under its own power. The wooden bridge across the narrows in Dunchurch had a section made of two-by-ten planks standing on edge that could be taken out so the tugboats could go through. As McCarl made his way through Dunchurch a crowd of women and children gathered on the west side of the road, and all the men around town were at Farley's blacksmith shop east of the bridge. Having an audience, McCarl stopped in the middle of the bridge to pose. Then he gave her steam ahead. This kicked some of the loose planks over, and the hind wheels broke through. McCarl stepped off the driver's platform and watched the tractor go backwards into the lake. Jim Sands, a long, lean fellow, had been holding a plank straight under one of the front wheels. As the outer end of it went up in the air, he hung on until he was about eleven feet long, then let go and fell back down wailing. The old men at the blacksmith shop ran inside and slammed the door, expecting the tractor to blow up.

So there was the new McDonald, lying on its side in the lake with the flywheel broken off and only the front showing above water. After it was jacked up, pulled out and fixed, Jack found it wouldn't go up the Grange Hill either. When the wheels started to spin he dropped off one wagon then another, until he was trying to make the hill with the engine alone. All he did was chew up the Grange Hill.

Then he bought a sawmill to run with it, and put it down at Wye's Lake, where it met more misadventures. The Vankoughnett boys, home on leave from the army, fired it up one Sunday and headed it backwards down to the lake to take on water. They couldn't stop it, and there it was, in the lake again. After finding it wouldn't run the mill very good, he took it back up to Dunchurch and put it in a shingle mill, where it worked fine until the mill burned and put it out of commission. So the McDonald went back to the company to be rebuilt, never to return.

JIM McARTHUR (b. 1886)

Jack Brownell came in to Bolger to cut illegally. That would be 1908. He'd been cutting out at Deer Lake in McKenzie and had to move; I guess Ritter [Sam Ritter, woods superintendent for Graves Bigwood & Co.] had got wise. There was more than Jack Brownell cut timber illegally, you know. He wasn't the only one. A firm in Brantford had turned out a portable mill, beautiful thing, gasoline powered. You made the framework and when the mill came you put it up. And you could move it quick. He was cutting hemlock, thousands of feet. All the houses around Bolger were built of hemlock Jack cut. We bought off him. His wife was sawyer, she ran the saw. She was a big woman. She could pick up Jack and you and I and throw us around. She had to run it because Jack never worked very much. He was an awful boozer. I remember him, when he was making a bit of money, going out to Whitestone to get some provisions and the horses bringing him home. She met him, jumped up on the wagon, grabbed the blacksnake whip, and you never seen a man get the pounding he got. I thought she'd kill him. Anyway, one morning we saw him taking his team of horses down the road, and some things in the wagon. He took the mill. About four days after, Ritter came in and asked where John Brownell was, and if we'd heard anything about this mill. Well, nobody knew. He'd scattered the sawdust around and covered it over with brush, and they couldn't find any trace of it. And he'd cut the lumber for all our cabins, $10 a thousand — delivered, hemlock, every board ten or twelve inches. But he got it out and away he went. I think they caught him afterwards up in McKenzie somewhere, where he started up again. But he was only taking hemlock, and the company didn't give a damn for hemlock, those were just cull. Until they got the pine out; then they were valuable.

Log-sorting jack in Parry Sound, 1898. — PARRY SOUND PUBLIC LIBRARY

MY FATHER

ROY WAINWRIGHT (b. 1918)

My father was a steam-feed sawyer for the Peter Lumber Company in Parry Sound. Peter's mill was what they call a gunshot feed. The carriage was attached to a piston in a cylinder the length of the carriage track, and when you admitted steam to it the carriage would go like a shot. Then he sawed in the Byng Inlet mill. (He and my mother went up on the *Northern Belle*.) It was one of the biggest mills on the Great Lakes. It had two double-cut bandsaws, two mills under one roof. The logs came up the jackladder between them and were kicked right and left to each carriage. Steam-feed sawyers made good money. My father had made a bit and wanted to go into business on his own. [He bought the Whitestone Hotel.]

ESTHER EINARSON (b. 1910)

My father [John McAmmond] was born in Ottawa. Our family owned the farm in Ottawa that's now the Experimental Farm, but they heard about this area with all the pine in it, and they were going to make a fortune in board pine, as they called it. So they came up to Maple Island in 1880 when my dad was eight years old. They logged. My dad had a lumber mill run by a waterwheel, a huge thing away over my head. The water came through a flume and it went round and round.

They had the first post office, Glenila, up the North Road on the west side, just this side of The Cataract. [McAmmond's place] was called the Halfway House, halfway between Parry Sound and McConkey. All the cadge teams and sleighs and wagons went through there. The front part was a log building; that was where they lived first. Then they built a big place at the back that was the dining area, full of tables. Another area was the kitchen. There were rooms upstairs in the front, and all over the two-storey frame building at the back. I think there was room for maybe twenty-five.

When I was a kid there were the remains of a log building at The Wildcat. I asked my father once why they called that hotel The Wildcat and he said, "Because that's what it was: the wildest place in Parry Sound District."

JACK LAIRD (b. 1892)

My dad ran a gang saw at Byng Inlet. It was Graves and Bigwood then. I remember Bigwood as a great big handsome-looking man. We were there four or five years. It would be about 1904 we left.

Dad used to walk up [from Parry Sound] in the spring of the year, before the boats were running, with two or three teams of horses. It would take them two days; it was just a trail. The family went up on the *Northern Belle*. It took practically a day to get there by boat. One winter we stayed up there, and we had the best skating in the world. All the lumber piles went into the water where the slips were, and the water would come up through the piles and flood the rink for us. We were madder than heck because the boat came in and broke up the ice.

Hauling lumber for the Lake Rosseau Lumber Company. The front sleigh carries over 10,000 board feet. — BARBARA PATERSON

Paul Leushner sawing in his watermill at Waubaunik, c.1900. — DEL LEUSHNER

HOW I GOT OUT OF SCHOOL

GEORGE DOBBS (b. 1905)

How I got out of school, I had an ulcerated tooth and a big lump an the side of my jaw that went down into my neck. Miss Kramer came down and said, "What's the trouble with you?" I said, "I think I've got an ulcerated tooth." She said, "I think you've got the mumps. You'd better go home and not give it to the other kids." Boy, that was just up my alley! I was supposed to try the [high school] entrance [examination], but I knew I wasn't going to pass, so I didn't have anything to lose. I went over to Hosick's mill and they were just getting ready to start. John Hosick said, "How would you like a job feeding the jackladder this summer?" I said, "Right up my alley!" That was 1921. I got $2 a day. I worked all summer and had $50 at Exhibition time. I thought I had all the money I'd ever need. About $50, and I spent $20 at the Exhibition.

It was a dandy job. I just wore running shoes, or bare feet sometimes. When the fellow on the log deck started the chain I'd keep poking the logs on. I had to sort the logs, take a run of hemlock till I couldn't get any more, then swing over and take a run of basswood or spruce or pine. They had little [lumber] cars they pushed on a tramway, and you'd try to send up enough to get a car complete, so they wouldn't get the lumber mixed.

I worked there two years, then went to Boakview to work for Simpson and Beagan. I drove an old horse there, drawing out timber blocks and whatnot. Where I got my big break was when I went to Hocken's after we finished the mill at Boakview. They had a cut of logs that they didn't get [sawed], so they came down and hired the Simpson and Beagan gang, and we went up and put that cut out. It was a winter mill, they ran it in the wintertime. They had a hot pond. Steam from the engine went into a pond with a jackladder out of it. The engineer, Johnny Woolman, took a liking to me, I guess, and I worked for him. Next year I went [with him] to West River on the Algoma Eastern Railway in from Espanola, and when the machinist quit he went machinist and I took over the boilers. Boy, I was getting $4 a day. I was one of the chief guys.

JACK LAIRD (b. 1892)

I think I was twelve years old when I first started to work in the lathe mill. They had a big picket mill, or box factory, between the two sawmills, and there were elevators from one sawmill to the other, then to the burner. I used to pick the material that would make the pickets and the lathe out of that big chain. I tailed the edger in the regular mill the next year. There was a chain at the tail of the edger that [carried the boards] up to the trimmers, through those saws to the culler, and out to the end of the long chains to the trucks. It was mostly pine, with the odd bit of hemlock. The trims off the edger would be so damn brittle they'd be in a half a dozen pieces by the time it would get to the yard. Pine you could flip off, but hemlock you couldn't. We hated hemlock.

We worked eleven hours at night and ten in the daytime. They had electric lights. They ran the boilers off the edgings and other [waste]. They had conveyor chains that carried sawdust into the boilers. They had a big burner outside where they used to burn stuff that they didn't use for the mill. That burner, you could drive a team of horses in the darn thing. It would be maybe a hundred feet high, all steel, with a screen on top. A chain went up from the mill and dumped it in up high.

In the mill I got paid $1 a day for ten hours. The men were getting $1.75. We went for more and they said they couldn't afford to give it to us, and Dad says, "Go home." So we went home, and it took three men to do the job that two kids were doing. There were a lot of Englishmen around the mill in that day, Englishmen and Americans. There were a lot of guys from Michigan worked in the yard and around the mill, piling lumber and that sort of thing.

Most of the lumber went to the States by barge, a couple of big barges towed by a steam barge. They had lumber yards by the mile. There was a bottom tramway and a top tramway. The one tram ran out so far, then there were slips the boats came up in. The barges ran right up in the slips and the lumber was on both sides of them. They would run this pile so far, then they'd run that pile so far. They didn't have to carry it, it all went with rollers from the piles to the boats. It was re-piled on the boat. I may be exaggerating, but I imagine there would have been four or five miles of [elevated] tramway. The tramway was on piles and the lumber was on piles. There were little canals up between each pile of lumber.

There was a big office and store right opposite the dock. There was a main office where all the officials stayed, and there were four or five big dormitories where the men stayed; the men slept in dormitories and ate in one big feeding place. I imagine there would be a thousand men, roughly, working in the whole operation, just in the mill alone.

There wasn't too much done by hand even in those days. They fired the boilers with sawdust with overhead chains. They [used] some slabs, but there were only two or three men on a half dozen boilers. They had a big flywheel about twelve feet high and a belt about four or six feet wide, and that big thing would turn about 400 [revolutions per minute] to run all the machinery. There were very few belts in the mill, mostly chains. The jackladder chain went right into the water, and the logs would go up there six or eight feet apart, all day long. The carriage could take the logs that fast, and some were as far through as I was tall. They put over 4,000 logs through the two mills in a day, if I remember right. From where the mill was, up to the rapids, would be solid with logs. I've seen logs out there, honestly, two or three guys could get on it and couldn't birl it, probably six feet in diameter. I've seen deal, we used to call it, cut special out of the big logs, the damn stuff would be eight and ten inches thick, eighteen to twenty inches wide and probably sixty feet long. We used to have to run it at an angle to get it through the mill. That wasn't an ordinary day's milling — special cases, somebody wanted some special timbers. That stuff was white as snow, that pine, you couldn't find a knot if you looked for it.

There were two separate mills, two separate jackladders, and the lathe mill was between the two of them. One mill had bandsaws and the other a circular saw. The last year we were there, they took the circular saw off and put bandsaws in it too. Instead of the carriage going back and forth, the log just went through the four saws — two cut the slabs off the outside and two cut the "edger" to make the width of board they wanted, then kicked it over to the gangsaws. The slabs went on a conveyor that took them to the burner or the boilers, and the [lumber] came through to the edger.

That mill had two gangsaws, one that would take about twelve logs in one cut and the other would take six or seven. The saws ran up and down; every time the saw would come down the log would go through a half inch or so. Maybe twenty-five or thirty saws. They could adjust them for whatever width they were cutting. When [the boards] dumped out of the gangsaws, the ones cut from the centre didn't need to be edged and they went straight to the trimmers, where they cut the ends off them. [Those with bark on] went through an edger to take the edges off. There were two of us on the edger; one had to take the edges off the boards that came from the bandsaw and the other guy took the boards from the gang. From the trimmers [the boards went] to a culler outside, and from there on the chains to where the lumber [was loaded on wagons]. A log never stopped from the time it left the water till it was cut.

There would have been between fifty and seventy-five horses there. They used just single horses, no teams on the trams. One horse took one load. When they loaded at the mill they ran [a wagon] up to the chains, [and when it was full] they pulled it down and a guy took it away. They'd unhook the wagon [at the lumber pile], the front would drop down, and the [pilers] would pull the lumber downhill off the load. When he unhooked he took the [front] wheels back with him for another load.

The green lumber from the mill all came out on the top tram, so it was all downhill to pile it. It would stay there a year before they would ship it, so it was fairly dry.

The reason the night shift worked an extra hour was that they had a short shift on Saturday night; they got off three or four hours early on Sunday morning. You didn't get anything for nothing in those days. The eleven-hour shift got off at four o'clock Sunday

Inside the Holland and Emery mill at Byng Inlet, about 1900. — JAMES T. EMERY

Sawdust-fed boilers in the Holland and Emery mill.
— JAMES T. EMERY

A lumber schooner waiting to load at Byng Inlet about 1900. — JAMES T. EMERY

Steamer MOHEGAN *towing loaded lumber schooners out of Byng Inlet.* — JAMES T. EMERY

morning so the mills wouldn't be running when the people were going to church. The mills were closed from four o'clock Sunday morning until six o'clock Monday morning. Other than that they worked twenty-four hours. Well, they stopped for meals.

We got paid once a month. You worked two months before you got a pay. The married men and people that had homes, and I guess the men in the bunkhouse, they had little books that they'd tear out slips [worth] a dime, a quarter or a dollar. A book with stamps in it, that was the cash. That was the way they paid their grocery bills. The company ran the mill and the store and everything. Across the river there was one little [non-company] store. They had a school on top of the hill, and they had a church, but the Catholic church was over on the Britt side, I think. The Americans that they had working there were most all single men, and they slept in these dormitories and ate in the big kitchen off the main part where the offices were. The Yankees brought baseball over and we had a lot of games. There was quite a big baseball field, all flat and covered with sawdust. Those Yankees were pretty good baseball players.

The mill shut down before it froze up and we always got out of there before the last boat went out. There wouldn't be but very few men around there in the wintertime. The only winter we stayed up there we didn't get out in time. I had typhoid — a lot of typhoid up there, no sewage systems. Mother boiled all our water. They had a big windmill to pump water; you had to go to the windmill to get your water. Everybody seemed to be healthy and happy — of course all they knew was work.

There was no way to get out; I don't think there was ever a road open out of there in winter. Whatever you had in the fall had to make it to spring, although the mail came in some way. I know a dogteam took the mail from Byng Inlet to French River. Besner, I think his name was, a Frenchman. He worked in the mill in the summer and drove the mail in the winter, about twenty-five miles from Byng Inlet to French River. He had three or four dogs on a sleigh.

Graves, Bigwood and Company lumber piles at Byng Inlet. — EVERETT KIRTON

Holland and Graves mill at Byng Inlet, c.1900. — JAMES ISBESTER

The Parry Sound Lumber Company's mill in 1900. — PARRY SOUND PUBLIC LIBRARY

BAD LUCK AND TROUBLE

BILL LITTLE (b. 1899)

Archie Perreault was on the repairs [for Schroeder Mills]; he looked after the cars, fixed them up. At the mill they had twelve or thirteen tracks in to the boardway where they loaded the lumber on. When a car was loaded, they took it out and put another one in. About ten o'clock in the morning one broke away, got past the yard engine and jumped the track. When Mark Nichols and I came in, the super asked us to hook on. We had the coach on, and two or three flatcars. We backed up and hooked on. I said to Mark Nichols, "I guess I'll take one of the dogs." You know what a dog is — that you put down on the ties so that the wheel will come up. You could spike them down, but we didn't have any spikes. I took the one up, and Archie Perreault went down and got the other one and put it on that side. The super yelled at us to stay back. You see, the lumber was piled up high on one side, maybe fifty planks. We should have chained it across, but we never thought of chaining anything. Archie says, "I'll take this stake (it was about six feet long) and hold that dog." I don't know, the Lord must have told me to step back. When they gave a lift and it hit the rail, the lumber came down and covered Archie and killed him. I felt sorry for him. When we dug him out he was dead. His wife had just had a baby in Parry Sound. He had come up on Number Twenty-seven that morning and walked the ten miles in to Lost Channel to go to work. I felt awful sorry about that.

ED PLETZER (b. 1887)

When I was twenty-six, the year I was married, I was canting in Vigrass's mill across the river from where I lived [in Spence Township]. Alex Reid was the sawyer. The first log that we put on one morning, he made a cut and was bringing the carriage back and he caught the dog. The guide pins to keep the saw straight had moved out and the saw was wobbling. The saw broke the block the dog runs in and fired the dog, and it hit me in the face. It broke my jaw here and here, and the top half was pretty near knocked right out. It knocked me down and I didn't know anything about it. The fellow that was running the edger saw it and came and turned me over or I guess I would have drowned in my own blood. They got me up and in the camp, and got me laying on an old bed tick, and the blood was running right down the floor. They got two doctors to come, one from Magnetawan and one from Sprucedale. The doctors couldn't do much but find out what was wrong with me. A fellow there said, "We'll send him to Toronto," and another one said, "He'll be dead by the time he gets there." "Well," the first man said, "if he goes to Toronto he has a chance; if you send him to Parry Sound they will kill him." And I believe they would have.

They took me to Seguin Falls in a democrat, with one fellow holding me up. They phoned the agent at Orrville and he said he would hold the train up for two and a half hours. I got to Seguin Falls in time to catch the train and went to Scotia Junction, and caught another train for Toronto. I was awake all the time. I got into Toronto at dark that night, and the doctor came down next morning to look me over — Dr. Bruce, a specialist. He looked me over and said, "He hasn't much blood left in him." I was in the hospital ninety days.

ARNOLD McDONALD (b. 1907)

Charlie Carter from Penetang was manager there [Holt's Mill]. I was pretty young, and he called me "Colonel." He said, "Colonel, I'll put somebody down here and you go and take the team on the dump cart." That poor bugger that went down over the bank, he hated to back up. That fall when the mill finished he kept me on there, yard team, shunting cars. They loaded cars with dry slabs. We had a place built up, and when you drove the wagon up you put a plank from the reach of the wagon to the floor of the car to walk on. I was helping them and had a big armful of slabs in front of me. When I went to step I missed the car floor and went down. There was the head of a spike sticking up in the floor of the car where my knee came down, and it went in my knee. My brother was barn boss, so he took my team and I went down to the office. It wasn't hurting, I guess it was numb. Charlie Carter was in the office, and Charlie Foley the clerk.

Soon as I pulled my pants up, Carter said, "By god, Colonel, you're going to the hospital." That was about eleven o'clock in the morning. I said, "Aw, no." He said, "I had trouble with a knee, and they're nothing to fool with." He said, "You go to the camp and we'll get you out of here." I walked in to dinner and while I was sitting there it stiffened up. You talk about pain. Holy Christ! The sweat ran off me. To take me out to Bolger they had a kind of arm-chair somebody had made out of a barrel, and the only way I could get any ease at all was to sit with my foot on a box and my other leg on top of that. Still the sweat was running. They put me in that on a flatcar — the size of me on a flatcar — and took me out to Bolger and loaded me in the baggage car of the train. They got me in there in that old rocking chair, and my feet propped up. I had never been in a hospital. It was the old Stone Memorial, and Dr. Denholme and Dr. Findlay were the two doctors. They told me I'd be out in four or five days, but blood poison set in and I was there five weeks. They carried me to the operating room three times.

JACK LAIRD (b. 1892)

The Parry Sound Company had a box factory over where the oil tanks are, you know that hill where you go up to the harbour. They had a bit of a canal about ten feet wide down to the bay that they used to float the stuff up. In 1906 I was tailing the saw for a guy shoving it through, and I got my finger too close and cut about an eighth of an inch off it. The blood started to fly and I flew too. Never went back. I started to work on the railroad. I worked from when I was twelve till I was eighty. Oh, I missed work the odd day, you know.

Loading lumber aboard THE SEGUIN, *Parry Sound.* — PARRY SOUND PUBLIC LIBRARY

EPILOGUE

ARTHUR MACFIE (b. 1893)
(From a letter written home in 1917 from a military hospital in the English countryside, where he was taken to recover from a wound received at Vimy Ridge, in France.)

There is a little dam below here, and when I hear the waterfall I think of driving days gone by. But I wasn't satisfied then, I had to get into the Big Drive in France.

The glory days of logging went into decline when the recruiting offices of the Great War drained off the young and daring, the class of men who imparted much colour and vitality. In the decade following the war the last remaining stands of old growth pine were rounded up and floated to sawmills, and the second wave of logging for species initially passed up, was well under way. By the beginning of the Second World War the era when timber was harvested exclusively with power generated by the muscles of men and horses, was already history.

Arthur Macfie, in foreground, on Shawanaga Lake. — MURIEL MACFIE

201

GLOSSARY

Alligator

An amphibious steam-powered vessel which pulled booms of logs across lakes with a cable and winch. Its manufacturer, West & Peachey of Simcoe, Ontario, called it a *warping tug,* but few river-drivers did.

Barber chair

If a leaning tree being sawed down broke off prematurely, it left a tapering slab resembling the back of a chair towering over a stump, which was then called a barber chair. "Barbering" a tree wasted timber and posed a threat to the sawyers who bungled the job of felling it.

Bag of logs   see *boom.*

Bind

When a saw cut closed on the saw and stalled it, it was said to bind. This could be forestalled by inserting a wedge.

Block

A sheathed pulley, usually made of iron. Also called a *sheave.*

Block of logs   see *boom.*

Board timber

A section of tree trunk which was hewed flat on two sides before removal from the bush.

Boom (boomstick, boom timber)

A long length of tree trunk. When a number were chained end to end with *boom chains* to encircle a mass of floating logs, booms, logs and all were commonly referred to as a *boom of logs,* or, under certain conditions, a *bag of logs* or a *block of logs.* Extra-thick *storm booms* corralled logs being towed on the Great Lakes. A *glance boom* fenced floating logs out of bays and shallows, and a *switch boom* served as a gate to stop logs from entering dams and rapids. A *flat boom* served as a floating walkway in front of a sawmill or dam. To *boom out* meant to catch logs at the foot of a dam or rapids and corral them in encircling booms for towing away.

Booyaw

An impromptu feast, usually of boiled fowl, sometimes stolen.

Bow gang

The men who worked at the front of a log drive.

Brag load (or show load)

An enormous sleigh-load of logs built for the fun of it. A *prize load* was the biggest regular load hauled in a season.

Breaking down

Loosening the front logs on a skidway and rolling them down to the loaders.

Buck beaver

The foreman of a road-building crew.

Bullrope

The rope with which a *bullroper* guided a log being hoisted by a jammer.

Bunkbound

A lopsided sleigh-load of logs pressed one end of the bunks down against the benches, making it hard for the horses to steer.

Cadge

To transport supplies by sleigh or wagon.

Canthook

A lever with a hinged hook for rolling logs (see illustration).

Capstan crib

A raft with a large spool mounted in the centre, with which horses or men wound up a rope to pull a boom of logs across a lake. River-drivers commonly corrupted it to *capsule* crib.

Capsule   see *capstan crib*

Check

Splitting of the ends of logs and boards caused by exposure to sunlight.

Chickadee

A boy assigned the job of keeping log-hauling roads level and the runner tracks clear of horse manure. A grown man doing the same job was more likely to be called a *giper*.

Chopper

The axeman in a log-making crew. He notched the trees for the sawyers, measured the fallen trunks into sawlog lengths, and lopped off the top.

Choreboy

He kept the lumber camp cookery and sleeping camps supplied with wood and water, tended fires and lights, and woke the men in the morning.

Cookee

A cook's helper who washed dishes, set tables and in most cases did some of the cooking.

Crazy wheel

A cable and drum affair which controlled a loaded log sleigh's descent down a steep hill.

Crib

A large raft made of squared timbers.

Culler   see *scaler*

Cut a log

To control the course of a rolling log by forcing one end back or forward.

Deadhead

A sawlog, one end of which is floating on the surface of the water and the other lying on the bottom.

Deck

To pile logs with a decking line.

Decking line

A pulley-equipped chain which, when pulled by horses, rolled logs onto a skidway or sleigh. It was also used to haul the barrel which filled a road-icing tank.

Depot

A centrally located headquarters of woods operations.

Dog

The name was applied to a variety of devices having in common an iron point which was driven into a log to hold it firmly or attach something to it.

Donkey engine

A small portable steam engine with an upright boiler.

Gipe

To patrol and apply minor maintenance to a log-haul road (see *chickadee*).

Go-back

A detour from the main log-haul used by an empty sleigh returning for another load.

Gunned log

A log which an improperly rigged decking line has caused to swing endwise, like a cannon barrel.

Handspike

A short pole employed as a pry.

Handyman

A semi-skilled craftsman who built sleighs and did carpentry around a lumber camp.

Head block

A log at the front of a skidway upon which the ends of the pair of skids forming its base rested.

Hew

To square the sides of a length of tree trunk with a broad axe.

Hooksman   See *roller*.

Jackladder

An endless chain which elevated logs from millpond to sawmill.

Jag

A part load.

Jam dog

This heavy iron hook, to which a rope pulled by horses or men was attached, was used to loosen individual logs in a log jam.

Jammer

An "A" frame rigged with cable and pulleys and used for slinging logs aboard a sleigh.

Job

To take out logs under contract.

Jumper

A small sleigh of simple construction.

Kerf

A cut made by a saw.

Knockdown

A wooden pin driven into a hole bored in a timber.

Loose

Floating logs that are not corralled within booms. To *run the loose*, that is to travel on foot over these logs, was more difficult than doing so on logs packed tightly together.

Patent plow

A factory-made snowplow fitted with rutters that gouged runner tracks in iced roads.

Pig's foot

A pair of these double-clawed hooks grasped logs for a jammer to hoist in the air.

Pointer

A large rowboat, pointed at both ends. In later years gasoline engines were installed in some of them.

Pung

A small horse-drawn sleigh which carried the noon meal to bush workers.

Pup

The hook which anchored the end of a decking line.

Raganooter

A fallen tree which lies partly in the water of a lake or river.

Rakers

These blunt teeth were placed at intervals among the cutting teeth of a crosscut saw to drag sawdust from the cut. Also called *drag teeth*.

Rampike

The skeleton of a long-dead but still-standing coniferous tree. Also called a *chicot*.

Roller

A man who piled logs on skidways by rolling them up with a canthook, or guiding their ascent if a decking line was used. Rollers, sometimes also called *hooksmen*, usually worked in pairs.

Sandpiper

A man posted at a hill who, by means of sand spread in the runner tracks, controlled the descent of loaded log sleighs. Also called a *sandman*.

Saving dam

A stop-log dam built to store water for the log drive.

Scaler

A person who measured logs or lumber to determine the volume of merchantable wood.

Score hack

To chop deep notches at intervals along the flank of a felled tree trunk to facilitate the job of hewing it into square timber.

Scraw

Tangled trees or brush through which it is difficult to walk.

Sender

He guided a log's ascent from skidway to sleigh. Where a decking line was used, his job was identical to that of a *roller* but when logs were raised with a *jammer*, he held a rope instead of a *canthook*.

Shanty

A lumber camp. Men who worked in lumber camps sometimes referred to it as "shanty-ing."

Shantyman

A career lumberjack.

Sheave   See *block.*

Shit out

Describes the predicament where logs roll out from beneath a pile, causing it to collapse.

Shoepac

A type of footwear having characteristics of both a boot and a moccasin.

Skid (verb)

To drag logs from where they were made, for piling in a central location.

Skid (noun)

One of a pair of tree trunks upon which logs were piled to make a *skidway*, and also the slanting poles up which logs were rolled to build a skidway or sleigh-load of logs.

Skidway

A trestle on which sawlogs were piled to await the winter sleigh haul to lake, river or saw-mill. Often used in reference to the pile of logs itself. A *fly skidway* was built on a steep downslope to take advantage of gravity.

Sloop

A crude sled, the runners of which were usually roughly hewed from curved tree trunks.

Snye

A short branch road.

Sorting jack

A floating gate where logs belonging to more than one company were separated.

Spill logs

To open a boom and allow the logs to float free.

Square timber

Sections of tree trunk hewed flat on four sides before removal from the bush. If less outer wood was removed, leaving round corners, it was called *waney timber.*

Spring out

To turn out at the end of a winter's work. Normally applied to horses being put out to pasture, but sometimes used in reference to lumberjacks in the idle season between the closing of lumber camps and the start of the river drives.

Spud

A chisel-like tool for separating hemlock bark from the trunk.

Stall

To put a horse off its feed by giving it too much.

Stub-shot

Extra length left on a sawlog to ensure full-length lumber could be sawed from it.

Swamp

To transport hemlock bark or cordwood from where it was cut to a point where it could be loaded on sleighs or wagons. Chopping the necessary trail was also called "swamp-ing."

Sweep

To roll stranded logs off the shore and prod them out of backwaters at the tail end of a river drive. Logs caught in an eddy were also *swept* out with booms chained end to end.

Time

Wages due.

Turned butt

Describes a sawlog which is markedly larger at one end than the other.

Turkey

A sack containing a worker's personal belongings.

Tail down

To roll logs down to the *senders* at the front of a skidway.

Tail gang
   The men who worked at the tail end of a log drive.
Tail-saw
   To receive lumber cut by a saw and direct its further progress through a sawmill.
Tank
   A sleigh-mounted watertight box which streamed water on haul roads to apply an ice surface.
Thousand
   A contraction of thousand board feet (one thousand square feet of board one inch thick), the standard measurement for both sawlogs and lumber.
Timber limit
   A tract of public land upon which a lumber company was licensed to cut timber.
Top-loader
   He was the nimble-footed architect who stood upon and built a sleigh-load of logs.
Trail-cutter
   He used an axe to cut brush and windfalls from the path of a skidding team. Sometimes called a *swamper*.
Trip hook
   A levered hook which could be released without relaxing tension on the chain to which it was attached. Also called a *shovel grab*.
Turnout
   A place for an empty sleigh to pull off a haul road to allow a loaded sleigh to pass.
Van
   A small stock of necessities kept for sale to the men in a lumber camp.
Walking boss
   The woods superintendent for a logging company. Often shortened to *walker*.
Waney timber   See *square timber*.
Withe
   A substitute for rope made by twisting the stem of a green sapling.
Wrapper
   A chain which bound a sleigh-load of logs in place.